EUREKA

*The Celestial Pattern at
Times of Historic Inspiration*

by Nicholas Kollerstrom Ph.D.

New Alchemy Press

Eureka

Published in 2012 by New Alchemy Press

www.newalchemypress.com

Copyright © 2012 Nicholas Kollerstrom M.A. Cantab, Ph.D.

The author has asserted his moral right to be
identified as the author of this work.

All rights reserved.

No part of this publication may be reproduced or utilised in any form or by any means, electronic or mechanical, including photocopying, recording or by any information storage and retrieval system without the prior permission of the publishers.

ISBN 9780-9572-79902

Enquiries should be addressed to the publisher.

Printed in Great Britain.

Charts & graphs by *Solar Fire Gold*: pub. Astrolabe Inc.
[http://www.alabe.com]
Harmograms: Mike O'Neill's *Harmogram* program
Graphics-editing: *Gimp*; Adobe's *Photoshop Elements*
LaTeX typesetting: *Kile*; P. Brachet's *Texmaker*

Front cover: The Eureka Moment — An Alchemical View, M. Maier,
Atalanta Fugiens, Oppenheim 1618 (*Wellcome Image Library*)

'A fundamental shift of outlook on the part of scientists might bring a healing relationship between science and valid aspects of mysticism, might open the frontiers between objective knowledge and subjective awareness.'

W. Beveridge, *Seeds of Discovery*, p.118.

Contents

Acknowledgements xiii
1   Introduction 1

I   Eurekaology

2   Quintiles, Septiles and Genius 19
3   The Act of Creation 33
4   The Angles of Inspiration 49
5   The Eureka Effect 63
6   The Moment of Illumination 81
7   Silhouette of the Septile 117

II   The Moment of Invention

8   Uranus the Awakener 129
9   When Inventions Worked 143
10   Structure of the Trine 221

III   Perspectives

11   Destiny and the Uranus Cycle 231
12   DNA and the Decile 261
13   The Synastry of Invention 283
14   Descartes' Dream 299
Appendices 315
Glossary 345
Select Bibliography 349
Index 353

Figures

1.1	Kekulé dreams the benzene ring	5
1.2	Trine aspect	9
1.3	Johannes Kepler	10
2.1	Kepler's Planetary Interval Geometry	20
2.2	Quintiles, septiles and trine aspects	23
2.3	Plutonium creation event chart	29
3.1	Arthur Koestler	33
3.2	The Eureka moment — an alchemical view	34
3.3	Louis Pasteur	40
3.4	The Andromeda Galaxy	42
4.1	Pentagons and pentagrams	53
4.2	Pentagrams and hidden pentagons	53
4.3	Golden Section/Spiral	54
4.4	Giza pyramids: septile angle	55
4.5	Dividing a circle into seven	58
4.6	The 'family' of heptagons	59
6.1	Kepler's Third Law Eureka moment 3rd harm.)	85
6.2	X-ray hand	95
6.3	Wilhelm Röntgen (Roentgen)	96
6.4	X-Ray event chart	97
6.5	X-ray e-moment 7th harm. & close-up	98
6.6	Penicillin e-moment 7th harm. & close-up	101
6.7	Laser e-moment 7th harm. & close-up	109
6.8	Fermat's Last Theorem — Solved event chart	111
7.1	Septile overtones 42nd-H	118
7.2	Orb of a septile: (a)natal group (b)e-moments	122
7.3	Edison's e-moment event chart (cnjs & opps)	124
7.4	Edison's e-moment 7th-H	125
8.1	Anniversaries of invention	138

8.2	I-moment scores: cnjs, opps and trines	140
9.1	Barometer event chart	145
9.2	Lightning conductor chart: kite formation	147
9.3	Lightning conductor event chart	148
9.4	Hot air balloon	150
9.5	First balloon flight event chart	151
9.6	Balloon over Paris event to Montgolfier natal	152
9.7	First balloon/balloon over Paris compared	153
9.8	Montgolfier portrait	154
9.9	Anaesthesia used in surgery	163
9.10	Anaesthesia event chart	165
9.11	Foucault's pendulum	167
9.12	Foucault's pendulum event chart	168
9.13	Electric light	173
9.14	The Wright brothers' first flight in Kitty Hawk	178
9.15	Powered flight event chart	180
9.16	Maiden flight, 1941	194
9.17	Jet plane event chart	195
9.18	Jet plane, showing Mars aspects	195
9.19	Jet plane lift-off 3rd harm. & close-up	196
9.20	Nuclear Power event chart	198
9.21	Digital computer program event chart	202
9.22	First program 2nd-H harmogram	204
9.23	Laser beam event chart	210
9.24	Lunar landing event chart	215
10.1	I-moments and major aspects to Uranus	222
10.2	I-moments and trines & septiles	225
11.1	Newton's prism experiment	235
11.2	Einstein's 1905 transits	243
11.3	Albert Einstein natal chart	244
11.4	Neptune-Uranus opposition of 1905 16th-H	246
11.5	Relativity Theory event chart 'mystic rectangle'	249
11.6	Darwin's Uranus-trine return chart	254
12.1	Tycho Brahe natal 7th-H	263
12.2	Kepler natal chart showing quintiles only	264
12.3	Johannes Kepler natal 5th-H	265
12.4	Quaternions event chart	268
12.5	Quaternions 4th-H harmogram	270

12.6	Seaborg graph: natal & Uranus/Pluto quintiles	272
12.7	Jupiter-Saturn square Pluto 4th-H	274
12.8	Plutonium creation and quintiles	275
12.9	DNA double helix	276
12.10	DNA 10th-H harmogram	277
12.11	The pentagon and the Golden Triangle	278
12.12	DNA 7th-H harmogram	279
13.1	T. H. Huxley natal chart	294
13.2	James Watson natal 7th-H harmogram	297
14.1	Descartes' dream 7th-H harmogram	304
B.1	Aspects Expected chart	323
B.2	I-moment and trine aspects	324
D.1	Lost and found Eureka moments	330
F.1	DNA discovery Gaussian (normal) curve	338
G.1	The US sci-fi novelist Philip K. Dick	339
G.2	Harmonic frequencies for 17 mystic moments	342

Tables

2.1	Orbs for septile, quintile & trine aspects	22
4.1	5th harmonic aspects	52
5.1	The Eureka moments (n=23)	64
5.2	Famous scientists (timed) quintile/septile scores	70
5.3	Excess of aspects	73
6.1	Synastry between Galle and Leverrier	91
6.2	Natal: Tombaugh Pluto Discovery Orb	105
7.1	Natal Group aspect frequencies	119
7.2	E-moment group 7th-H using Addey orbs	119
7.3	Natal group planetary frequencies (Q+S)	120
7.4	Event group planetary frequencies (Q+S)	121
8.1	Inventions: days when they first worked (n=36)	136
8.2	Inventions group (n=36): total uranus aspects	137
8.3	Strength of the Uranus aspects (for 5° orbs)	139
10.1	Celestial aspect frequencies for n=36 group	224
12.1	DNA E-moment: The seven 10th-H aspects	278
13.1	Kepler and Newton synastry	285
13.2	Wallis-Hobbes synastry mathematical dispute	287
G.1	The Dated Mystic Moments (n=17)	340
G.2	Astrological symbols	347

Acknowledgements

Thanks firstly to Mike O'Neill, co-creator of the Eureka Effect, without whom it would not exist; and to Claire Chandler, without whom this would still be a mere dusty manuscript; and to Derek Norcott who made it happen.

The present work has not reproduced text, tables or figures from the 1996 work 'The Eureka Effect', which was co-authored with Mike O'Neill. Therefore, I see no reason not to publish this work as a sole author. The responsibility for the content and for the decision to publish is mine alone.

CHAPTER 1

Introduction

THERE ARE MOMENTS OF ILLUMINATION, in the lives of scientists and inventors, when the key to a problem dawns. Something new appears. For the individual concerned, such a moment may appear, in retrospect, as the turning-point of their lifetime. Many of us have had something like such 'Aha!' experiences. The history of science is bejewelled with these celebrated moments when inspiration dawned, and these remain of perennial interest. These special moments will here be used to test a concept of celestial influence.

Archimedes was asked about a crown, by the tyrant of Syracuse. This led to the events of the primal, the original, eureka legend. He leapt out of the bath when a solution to the problem dawned upon him, and ran naked down the streets crying:

"Heureka, Heureka!"*

(The Greeks sounded the 'h'.) That is how the Roman writer Vitruvius described it, but unfortunately his account was two centuries after the event. More recent eureka-moments can however be firmly dated. In this treatise, we seek out such *times* when inspiration struck. Thereby we part company from all previous studies of the psychology of scientific inspiration. Our concern is not so much with 'what' a scientist realised, as 'when' he did so.

---

*The word 'Heureka' is the past participle of 'Heuro' meaning 'I find'; as in 'heuristic', which means reasoning leading on from what has been found. The epsilon, first letter in the Greek word, is aspirated, which means that the 'h' is sounded (thanks to James C. for explaining this).

We allude to these as *genesis-moments*, as times when something of historic significance first appeared. This term will also allude to those famous dates when notable inventions first started to operate: these are exciting moments of innovation of a more concrete and public type. Is there some special quality of time at such moments? At these times, when history takes a new turn, as could not have been predicted from the past, do the heavens show any special characteristics?

Does the solar system bear a particular configuration or harmony at these moments of illumination? Do certain types of aspect tend to be associated with such moments, and indeed do they stimulate such moments? Celestial aspects are angles formed between the planets, around the Earth. The 'permitted' angles here are those formed by simple whole-number divisions of the circle. In attempting to answer these questions, our inquiry turns out to provide some definite evidence for the effective working of celestial aspects — maybe, for the first time.

The celestial aspects connected with the prime numbers five and seven are known as quintiles and septiles. These aspects, which come from dividing the circle of the zodiac into either five or seven equal parts, have been little used until recently, say a few decades ago. Quintiles have come to be associated with creative ability and with one's 'art', while the septiles are credited with having an inspirational quality, as seems appropriate for a eureka-moment.

At these times of altered awareness, is the individual somehow in tune with the *Makrocosmos*, that is the pattern of the solar system as a whole? There is an irony here, in that scientists have long been sceptical towards any such notions, whereas we are using the timing of their achievements to validate this concept. My impression is that it would be harder to conduct such an inquiry using the history of art (say), because such decisive and innovating moments would be harder to find.

Our study concerns a type of exaltation, which often appears as the thrill of a lifetime, whereby an individual comes to be placed in a special privileged relation to the world, of becoming a focal point for its progress. Arthur Koestler's classic work, *The Act of Creation* proposed his key notion of 'bisociation' for those times when creative synthesis occurs. We use his notion to define what is meant by these exceptional, once-in-a-lifetime experiences. Different patterns of logic fused together at these moments, giving a new way of looking at things. One cannot expect these moments of mental birth to be fully definable through terms derived from ordinary life. Perhaps, in some celestial ante-chamber, Koestler's spirit might even approve of the cosmic viewpoint on the *Act of Creation* here attempted.

We are concerned with the *subjective experience* of the scientist, the very part surgically removed from the final report, in order to locate the 'genesis moments'. Tragically, both the emotional experience of the discovery and the time when it occurred are so often lost irrevocably, due to the tradition of objectivity or 'glacial remoteness' in a scientific report. All too often the experience is only mentioned two or three decades after the scientist has retired, by which time the event may have acquired a rosy hue in memory. For example, Albert Einstein had some fine eureka experiences, but alas only recalled that for his *Theory of Relativity* (while walking in the Swiss alps, near Berne, with his friend Besso, in 1905) *seventeen years* after it dawned upon him, and once his theory had become accepted (chapter 11). Nonetheless, no one seems to have doubted the veracity of his memory.

We have been conditioned to regard the realms of science and technology as being very much opposed to anything associated with astrology. But in fact, the irreversible progress of our scientific era provides the very conditions necessary for testing of astrological notions. For example, when James Watson discovered the DNA structure, nothing like that had ever happened before. It was a new beginning, a genesis-moment, *for which reason* the theory of astrology should be applicable to it. This work examines the astral context of such seed-beginnings.

Our inquiry begins with the locating of such dates, of documented E-moments (as we'll call them). Such events formed a turning-point in the life of the scientist as well as affecting the course of history. An example would be Dmitri Mendeleef envisaging his Periodic Table of Elements. Mendeleef, in St Petersburg, had made cards for each of the known elements and shuffled them about on the carpet, before dozing off. He had postponed a visit to a factory due that afternoon as if sensing that something important was due to happen. On awakening, he was able to assemble them into a new kind of sevenfold pattern. We can regard this event from different viewpoints: psychological, chemical, cosmic. Mendeleef's Table predicted the existence of elements not yet discovered and opened up a new perspective on the nature of matter. We should expect that some significant cosmic pattern would then exist. The Periodic Table revealed a sevenfold structure to the elements, but it was not *simply* sevenfold as an earlier theorist had supposed (John Newlands). Its seven tiers were to be echoed in the next century by a sevenfold model of atomic 'shells', as these would be used to explain Mendeleef's Periodic Table.

The eureka-ologist needs a modest acquaintance with the field of science in which the discovery is made, but only to a limited extent. The substance of what was discovered has only a secondary importance for us. For example, we needn't understand what came to Werner Heisenberg when he discovered his 'magic matrices' of quantum mechanics in 1925, in order to check out how that moment supports the hypothesis here tested. We study a link between the psyche of the scientist and the condition of the heavens. At the very least, this approach gives a fascinating new perspective to the history of science, which can otherwise be a rather arid affair.

Eureka insights don't just pop out of the blue. They only emerge after a long period of gestation and inner struggle, even verging on despair, when the sought-after resolution seems to be unattainable. We therefore need to study these special moments in a larger context, of the individual's life. Later sections put the 'moment of illumination' in the context of these larger life-cycles, looking at the overall shape of a lifetime in terms of the planetary cycles involved, and especially the main periods when creative work is achieved. Thus, Newton lived through a complete Uranus cycle of 84 years, and the stages of this cycle mark out with remarkable precision his main periods of achievement. Kepler, in contrast, lived through two Saturn cycles and this appears as his dominant planet. The creative spark often appears around the age of 28–30 when major steps in the Saturn and Uranus cycles synchronise, and we focus on this. Biographers of the future may want to use these cycles for mapping out the development of a person's genius, for example the seven cycles of Jupiter that Freud lived through.

Chapter 1. Introduction

Lost & Mythic E-Moments

Some of the best eureka stories turn out to be mythical and constructed some years later: for example, that of Newton and the apple.[1] The legend — ah, the legend! — of the eureka scientist derives much from the falling apple watched by the young Isaac Newton around the year 1666. However scholars nowadays place Newton's understanding of the law of gravity nearly two decades later, in the early 1680s, and view it as a more gradual process.[2] The apple story was first narrated by Newton *six decades* after the alleged event, in the year before his death.[3] As we lack reliable birth data for Newton,[†] the issue doesn't greatly concern us.

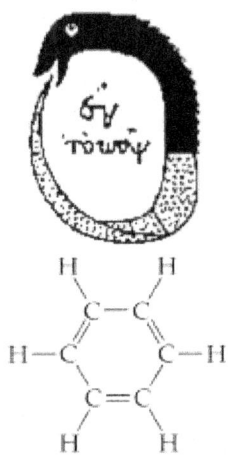

Figure 1.1: Kekulé dreams the benzene ring

Perhaps rather comparable to the falling apple story is the classic dream-vision of August Kekulé, on which doubt has recently been cast. Generations of chemistry students have heard how Kekulé discovered the structure of the benzene molecule, while dozing in front of his fireside in the year 1861. He was baffled by the enigma of benzene's structure (how could its six carbon atoms get to share three double bonds between them?) and was gazing into the flames in reverie, when lo! he saw the atoms joining up as in a dance, or like the Ouroboros, the serpent of mythology which eats its own tail: the ring structure of the benzene molecule! Kekulé published the structure in 1865, and the London Chemical Society proclaimed it to be 'the most brilliant piece of scientific production to be found in the whole of organic chemistry.'

It gave chemists an imaginative new outlook. But, recent scrutiny of his manuscripts over the crucial years has undermined the story, beloved by psychologists and students of creativity. Kekulé only related the dream-story *three decades* after it had supposedly happened

---

[†]There is a tradition of 2 a.m. on Christmas day for Newton's birth, but no source for it.

5

to him, right at the end of his life. Scholars now view it as brewed up in order to sidestep a rival priority claim.[4]

A well-described moment of illumination was that of the mathematician Henri Poincaré, while stepping onto a bus. In his magnum opus *The Act of Creation*, Arthur Koestler viewed this classic example as showing all the salient features of inspiration — its unexpectedness, as if erupting out of the blue from the unconscious, leading to the resolving of stresses that had built up for some time within the mind of the mathematician, stresses which could not be resolved along existing lines of thought:

> 'Contrary to my custom', recalled the mathematician, 'I drank black coffee and could not sleep. Ideas rose in crowds. I felt them collide until pairs interlocked, so to speak, making a stable combination. It seems in such cases that one is present at his own unconscious work, made partially perceptible to the over-excited consciousness ...'[5] [6]

The next morning, while stepping onto his bus, the nature of 'Fuchsian functions' suddenly dawned upon him, though he had previously felt sure that such a mathematical group could not exist. It seemed to Poincaré as if his 'subliminal self' had somehow reshuffled the options with which his mind had been grappling, to achieve a synthesis. As with the Mendeleef moment, we see the tension of previously unresolved endeavour, then reverie, then the sudden 'Aha!' as the new pattern clicked into place. Poincaré held forth at length about how he came by his insight, though neglecting to mention the small matter of its date, which means that this celebrated E-moment can only be dated within a week, in August of 1880.

Benjamin Franklin created the story of his kite experiment with lightning. He did this deliberately to give himself credit for a discovery, somewhat as Kekulé made up his story. It may be rather shocking for us to discover that these great mythical moments are really fictitious. Perhaps all that really concerns us here is that they cannot have a date, and therefore are not included in our survey.

Chapter 1. Introduction

From Conception to Birth

While assembling the eureka moments, cases turned up which seemed to resemble them but, upon inspection, turned out to be quite different. The book, *Eureka!* by Edward de Bono is subtitled, 'A History of Technology', and is merely a history of inventions. The word 'Eureka' means simply 'I have found it' and this could legitimately refer to the first working of an invention. Here we choose *not* to give it such a meaning, but to limit it to the inner lightning-flash of realisation. It took us a while to realise that two fundamentally different moments were here involved, sharing in common the thrill of innovation: one is an insight, the other a deed.

The second major category of event here investigated, is the invention or I-moment, when an invention first worked. Edison's light bulb first working, the first glimmer of Baird's TV screen in his Soho attic, the first atom bomb exploding, the first laser beam in California, the first holographic image — these thrilling (or terrible) events have given substance and direction to our concept of progress. We can be sure both of their innovative nature, and of their dates. Part Two looks at these. They are moments of achievement rather than insight, and because they are more public one finds much more books written about them. They were a lot easier to collect than the E-moments, and the group is larger. Britain has over the centuries been pre-eminent in both the I- and E-moments, showing (I believe) the creative and innovative nature of British culture.

Edison was the great inventor, but also seems to have had an E-moment, when the intuition of how to construct a phonograph came to him. It arrived while he was pondering the function of a Morse code tapper, and wondering how to make a record-player (phonograph). Some months later he constructed the apparatus, and declared that he was never so taken aback in his life as when his machine spoke to him. We classify Edison as a eureka scientist because of his documented E-moment, and thereby he bridges our two categories by also having a couple of (much better known) dated I-moments.

Only a few eureka experiences are on record in America, as we here use the term. As well as Edison's moment, there was one by Professor Charles Townes, when the principle of what was to become the laser dawned upon him in 1951, and finally — perhaps the best one, only described quite recently — was that which dawned upon astronomer Edwin Hubble, concerning the scale of the universe, after a night-time session at the Mount Palomar telescope in California. The eureka experience appears as mainly a European phenomenon! On the other hand, what America lacks in E-moments it has made up for in I-moments, producing more of these in the modern age than any other nation.

Two of our best-timed events concern the laser beam, both its E- and its I-moment. The former is the time of inspiration, which occurred when Charles Townes was sitting on a park bench in Washington D.C. early on a beautiful spring morning, admiring the azaleas. The I-moment happened a decade later when the invention was switched on, and this happened in Malibu, California, where Theodore Maiman in 1960 built his ruby laser. I actually corresponded with these two, scientist and inventor, whereby the date and the exact time of day of these two events were found. Also, both birth times were reliably obtained. That brings them into a marvellous focus and enables us to gain more detailed information about them. It gives a clear horizon position, against which the cosmos wheels every 24 hours — a thing lost for most of this study as such accuracy of timing is simply unavailable. These two types of moment, relating to head and hand, inspiration and perspiration, imagination and concrete reality, conception and birth, require different structures of interpretation, although they do have something in common. Part Two addresses the I-moments.

We will here show that some influence affected *when* the E- and I-moments tended to happen, as made some times less suitable for them, and others more. The characteristics of these two types of event differ in quality, for example the influence of the planet Uranus turns up quite distinctively in the I-moments. From comparing these two groups we get to look at the very shape of a celestial aspect, as has not been done before.

Chapter 1. Introduction

Kepler's View

> ... 'theorising' comes from **Theoria**, again a word of Orphic origin, meaning a state of fervent contemplation and participation in the sacred rites (**thea** spectacle, **theoris** spectator, audience). Contemplation of the 'divine dance of numbers' which held both the secrets of music and of the celestial motions became the link in the mystic union between human thought and the **anima mundi**. Its perfect symbol was the Harmony of the Spheres — the Pythagorean Scale, whose musical intervals corresponded to the intervals between the planetary orbits; it went on reverberating through 'soft stillness and the night' right into the poetry of the Elizabethans, and into the astronomy of Kepler.
>
> Koestler, The Act of Creation, p.260.

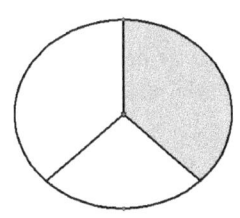

Figure 1.2: Trine aspect

By aspects, an astrologer refers to angles formed in the sky between planets in their journey round the zodiac. Whole number divisions of the circle of the zodiac form these important angles, for example a 'trine' is produced by a three-way division to give 120° angles between the planets, with respect to the Earth (figure 1.2). At the Full Moon, the Sun and Moon are directly opposite each other, i.e. 180° apart, and so they form an aspect of 'opposition.'

A musical analogy is here quite traditional:

> Students of astrology will be aware that aspects are most effective when exact, but that they possess an 'orb' of a few degrees each way within which their efficacy can still be felt. Beyond the orb there is a dead zone until the influence of the next aspect begins. One can regard the aspects, then, as marking a limited number of nodes of influence that activate the planets as they enter them. In the same way, one could call the intervals of music just such a limited series of nodes, also possessing a certain power, even if not exact.[7]

The present work will describe evidence for the working of celestial aspects. Evidence for their efficacy in human affairs remains controversial.[8] Chapter 4 discusses the celestial aspects which are of particular concern to our survey.

Of the scientists here discussed, only the three earliest ones, Tycho Brahe, Kepler and Galileo, might have been sympathetic to the present enterprise. They were employed as *mathematicus*, and a one-word modern equivalent to that term would probably be 'astrologer'. What was then called mathematics embraced astrology and astronomy, and was linked with music theory, as well as arithmetic and geometry. It is less easy to say what Galileo believed, since his views on the matter have been so censored by the official myth-makers, concerned to present him as 'modern'.[9] It is clear, however, that he was a practising, professional astrologer employed by the Medici family in Florence to cast charts, although he had to tone down this side of his work in later years as it became somewhat politically dangerous. Scores of books about 'Galileo' give no hint about this side of his activities.

Figure 1.3: Johannes Kepler (*Wellcome Image Library*)

Alone in the annals of modern science, Johannes Kepler not only practised astrology, but formulated with deep insight the conditions under which it worked and the extent to which it did so. He put forward a coherent explanation as to how it worked, insofar as these things can be explained.[10] [11] It is appropriate that we should follow his guidance on this matter, though readers seeking for more information on Kepler's views may experience difficulty in finding material in the English language. An important Kepler treatise on this topic appeared in 1601, as the foreword to an almanack for the coming year, entitled *De Fundamentis Astrologiae Certioribus*, which has been trans-

Chapter 1. Introduction

lated as 'On giving Astrology Sounder Foundations'. It was an astronomer's view on the extent to which the old notions of astrology were justifiable, and which needed to be thrown out. Before the gaze of astronomers had been obscured by the telescope and just after the death of Tycho Brahe, Johann Kepler asserted that:

> almost every motion of the body or soul or its transition to a new state occurs at a moment when the figure of the heavens corresponds to its birth figure ...[12]

He was discussing the fate of eminent persons of the realm, in relation to some celestial aspects which he foresaw that year. The Earth as a whole responded to the angles formed by the heavenly bodies, since 'Earth has a vegetative animal force, having some sense of geometry.' As a shepherd is pleased by the piping of a flute, Kepler explained, without understanding the theory of musical harmony, so likewise would Earth respond to the angles and aspects made by the heavens, though not in any conscious manner. Earth, he averred, was a sentient but not conscious being.

As an example of this, we may cite how what came to be called 'Kepler's star' appeared as exactly conjunct his descendant.‡ As a galactic supernova, it was the second to be recorded in Europe since antiquity, appearing in the year 1604 as a brilliant new star, on the zodiac and right next to a triple Saturn–Jupiter conjunction then in process.[13] Kepler was expected to interpret it, and it has been named after him ever since. It was closely conjunct the Galactic Centre, and being on his descendant, was directly opposite his Neptune rising in his natal chart. Such an inter-relationship illustrates the above quotation.

His work has not been much cited by astrologers, possibly because of a degree of scepticism he maintained concerning the principles of their craft — the zodiac and houses — leaving at times little more than the bare Pythagorean harmonies of the angular patterns.[14] Kepler introduced some new aspects, principally the quintile, formed by dividing the circle into five. He determined the efficacy of aspects from whether the ratios they formed by their division of the circle

---

‡Kepler's Descendant was 23° Sagittarius and Kepler's star was 22° Sagittarius, and the Galactic Centre was at 21° 35′ Sagittarius. When it appeared, on 17th October, Jupiter was at 20° Sagittarius, opposite his Neptune at 23° 10′ Gemini.

— such as 1:4 or 2:3 for the quintile and biquintile — were musically appealing, and also from the degree of symmetry of the regular polygons. Kepler's quintile aspects feature prominently in our study.

His great work, *Harmonice Mundi*, recently translated into English for the first time after three-and-a-half centuries, contains Kepler's third law of motion. However, as well as this astronomical principle, it also contains a vital discussion of astrological matters, i.e. it attempted to view these two different realms as some kind of unity.

The reason for the influence of transits was there explained as follows:

> The soul of the newly-born baby is marked for life by the pattern of the stars at the moment it comes into the world, unconsciously remembers it, and remains sensitive to the return of configurations of a similar kind.[15]

One ought to take the same interest towards Kepler's three laws of planetary motion, as to the three new aspects he invented. It is the very fault-line of schizophrenia in our civilisation, that one is expected to take an interest either in one or the other. Astronomers use his three laws in a physical science and astrologers use his three aspects in their intuitive art, as if outer and inner worlds had to remain quite unconnected. Kepler pleaded to his fellow astronomers not to 'throw out the baby with the bathwater' by rejecting the core of rational truth in astrology; but, they did so. Following Kepler's advice, we will hope not to be accused of trying to explain too much by astrology, or too little.

The dominant influences in Kepler's chart were Neptune on his Ascendant plus Uranus conjunct his Sun:[16] one wonders how he managed to understand himself without access to this vital information! We will mainly stay within the framework he laid down, i.e. our approach to matters celestial will be basically Keplerian. Our study deals with great men — and indeed a great woman, Lise Meitner — who laboured to construct a physical science, free of superstition or make-believe. It is only fair to their memory to take a somewhat minimalist approach towards astrology if one is to apply it to them. Kepler called himself a 'Lutheran' astrologer in throwing out the excess, and his treatise *De Fundamentis* concludes,

This completes what I think one may state and defend on physical grounds concerning the foundations of Astrology ...[17]

A Baconian Approach

Eureka-ology requires a historical approach, in tracking down those elusive moments. There were some fine E-moments at the dawn of the seventeenth century, when modern science was just stirring. This was a period before the Great Divide, between inner and outer experience, when astronomy and astrology were still being studied by the same people. Thus the Lord Chancellor of England, Sir Francis Bacon, when composing his magnum opus *The Advancement of Learning*, did naturally enough, allude to the major celestial event which had just occurred, a triple conjunction between Jupiter and Saturn:

> ... a conjunction like unto that of the two highest planets, Saturn, the planet of rest and contemplation, and Jupiter, the planet of civil society and action.[18]

He was calling for contemplation and action to be more linked together, here contrasting the introvert and extravert qualities traditionally associated with these spheres.[19]

Bacon wanted to see a sobered-down astrology, an *astrologia sana*. Astronomy, he complained, had become sundered from natural philosophy to its detriment, and had allied itself merely with mathematics, which had rendered it 'sterile'. The world stood in need, he said (in *De Augmentis Scientarum*, Bk. 3, Ch. 4), of a very different astronomy,

> ... a living astronomy, an astronomy which should set forth the nature, the motion, and the influences of the heavenly bodies as they really are.[20]

He regarded it as 'certain' (*quod nobis pro certo constet*) 'that the heavenly bodies have in themselves some other influences beyond heat and light.' However, exactly what these were lay hidden in the depths of physical nature (*intima Physica latent*), and it would require research programmes to discern them.[21] The 'reformed astrology' he was advocating would sift out traditions, and discard those 'as manifestly oppugne Physical Reasons.' He recommended investigating 'civil affairs' from a historical perspective to discern that which he was advocating: a sobered-down astrology/astronomy that would be truly useful to human affairs.

This forthright commendation by the founder of British philosophy, the very spokesperson of the scientific method, has been greatly overlooked. He did advise throwing out much of traditional lore, but accepted the influence of aspects. The restoration which Bacon advocated for astrology was a research programme of several components, including for example weather observations plus 'past experiments.' These involved the comparing of past events with celestial phenomena, just as here attempted.

Not being focused upon natal astrology, of which Bacon seems to have disapproved, this work may claim to fall within these guidelines. We proceed in the spirit of Baconian inquiry to ascertain whether or not any particular celestial aspect has an above-average frequency at the key moments. The philosopher Lord Bacon recommended that the principles of astrology should be tested by comparing major social events and revolutions with concurrent celestial phenomena,[22] and we attempt here to do that.

References

1. I. Bernard Cohen, *The Newtonian Revolution*, 1980, p.249.

2. http://www.dioi.org/kn/newton-gravity.htm N.K., 'Halley's Comet and Newton's late apprehension of the law of gravity', *Annals of Science*, 1999, p.56, pp.331–356. http://www.dioi.org/kn/newton-bio.pdf N.K., 'Newton, Isaac' in Biographical Encyclopaedia of Astronomers, Ed., Thomas Hockey, 2007, pp.830–832.

3. R. Westfall, *Never at Rest, a Biography of Isaac Newton*, 1984, p.154.

4. John Wotiz and S. Rudofsky, 'Kekulé's Dreams: fact or fiction? *Chemistry in Britain* August 1984, pp.720–723; J. Wotiz, Ed., *The Kekulé Riddle, A Challenge for Chemists and Psychologists* Illinois, 1997.

5. Jacques Hadamard, *The Psychology of Invention in the Mathematical Field*, Princeton 1945, p.14.

6. Arthur Koestler, *The Act of Creation*, 1964, p.212.

References

7. Joscelyn Godwin, *Harmonies of Heaven and Earth*, 1987, p.150.

8. The late Michel Gauquelin remained sceptical over evidence for aspects, see e.g., M. Gauquelin, *Written in the Stars, The Proven Link Between Astrology and Destiny*, 1988, p.253. I disagreed: N.K., 'Investigating Aspects', in *Research Methods in Astrology*, Ed. Mark Pottenger, ISAR, California 1995, pp.287–302.

9. N. Campion and N.K., 'Galileo's Astrology' special issue of *Culture and Cosmos* Summer 2003 Bath Spa http://www.dioi.org/kn/galileo/index.htm; Darrell Rutkin, 'Galileo Astrologer' in *Galilaeana* journal of Galilean studies Florence II, 2005. http://www.skyscript.co.uk/galast.html.

10. N.K, 'Kepler's Belief in Astrology', Ch. 5 of *History and Astrology, Clio and Urania Confer*, Ed., Kitson, 1989. http://www.skyscript.co.uk/kepler2.html, also http://www.dioi.org/kn/galileo/index.htm.

11. Vickers, Ed., *Occult and Scientific Mentalities in the Renaissance*, CUP 1984, Chs. 5,7,8.

12. J. Kepler, *De Fundamentis Astrologia Certioribus*, Prague 1602, translation in J. V. Field, 'A Lutheran Astrologer: Johannes Kepler', (Archives for the History of the Exact Sciences, 1984, Vol. 31 pp.190–268), entitled 'On Giving Astrology Sounder Foundations'; N.B. an alternative translation of his work by Prof. Bruce Brackenridge and Mary Ann Rossi as 'On the More Certain Fundamentals of Astrology' had earlier appeared (Proc. Amer. Phil. Soc. 1979, p.123): Field and Brackenridge had a row while collaborating on it and published separately.

13. N. Davidson, *Astronomy and the Imagination*, 1985, p.167, http://www.dioi.org/kn/NewStar.pdf.

14. See comments of Prof. Ken Negus in Foreword to *Kepler's Astrology, Excerpts*, Princeton, 1987 (ref. 12). For a more 'academic' view, see the essays by Rosen and Field in refs. 11 & 12.

15. Kepler, *Harmonice Mundi* 1618, Chapter 7; English translation 'The Harmony of the World' by E.J. Aiton, A.M. Duncan and J.V. Field, American Phil. Soc. Philadelphia, 1997.

16. N.K., 'Kepler's Chart', A.J., Nov. 1996, pp.371–7.

17. Kepler (ref. 12) Ed., Field.

18. 'Francis Bacon', Ed. A. Johnson, 1965, p.45 (Book One of *The Advancement of Learning*).

19. For validation of this Jupiter-Saturn polarity in psychological terms, see M. & F. Gauquelin & Sybil Eysenck, 'Personality and Position of Planets at Birth: An empirical study' *Brit. Jnl. of Soc. & Clinical Psychology*, 1979, 18, pp.71–5.

20. *'substantiam et motum et influxum coelestium'* The translation of this sentence is as given by Lord Macaulay in his 1837 essay on Lord Bacon, in his *Miscellaneous Essays* (Collins), p.419, there being no English translation of Bacon's *De Augmentis Scientarum*.

21. See Darrell Rutkin, *Astrology, Natural Philosophy and the History of Science, c.1250–1700* Ph.D at Indiana U., 2002. Quotes are from Bacon's *De Augmentiis Scientarum*, composed in 1623 after a lifetime at the hub of public affairs.

22. For more on Bacon's attitude towards astrology, see Patrick Curry, *Prophecy and Power, Astrology in Early Modern England*, 1989, p.61.

Part I

Eurekaology

(*Wellcome Image Library*)

CHAPTER 2

Quintiles, Septiles and Genius

'... yet the charts of creative people very often seem devoid of anything really noteworthy unless the fifth and seventh series of aspects are observed.'

John Addey[1]

THERE WAS A MOMENT OF illumination to which Kepler always attached a great importance. He was standing in front of a school blackboard facing a rather empty room of mathematics students, in a place called Graz in Austria, as one of the last Protestants in a Catholic city, in the year 1595, when a realisation dawned upon him. 'It will never be possible for me to describe in words the enjoyment which I have drawn from my discovery' he remarked some years later.

As Carl Sagan described the memorable moment:

> And one pleasant summer afternoon, deep in the interstices of one of his interminable lectures, [Kepler] was visited by a revelation that was to alter radically the future of astronomy. Perhaps he stopped in mid-sentence. His inattentive students, longing for the end of the day, took little notice, I suspect, of the historic moment.[2]

What then came to Kepler was a theory whereby the five regular Platonic solids were made to nest inside the six celestial spheres, of the planets (figure 2.1). Thereby he could explain the relative distances between the planetary orbits, from the Sun. This odd but charming notion seems to have fitted the data tolerably well. His first book, *Mysterium Cosmographicum* of 1596 gave expression to the grand scheme.³

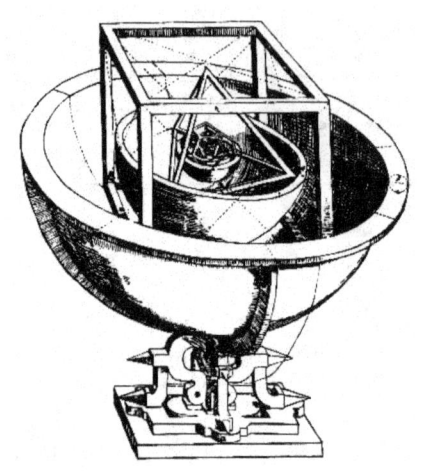

Figure 2.1: Kepler's Planetary Interval Geometry

In one sense, this was a classic eureka moment. That first publication gained Kepler recognition and brought his name to the attention of Tycho Brahe, then the Imperial Mathematician at Prague, yet its view of the Platonic solids nestling inside the planetary orbits and defining their relative sizes cannot be taken seriously. It is omitted from our E-moment list only for one reason, that no-one after Kepler ever took the idea seriously, i.e. it was *not true*.

Kepler felt that the Platonic solids with their different angles helped him to understand how astrology worked, and so this image, which remained important to him throughout his life, helped him to link together his trade of astrologer with his hobby of astronomy.

Kepler argued that his scientific and astronomical discoveries did not have an astrological causation. In a mood to oppose astrological determinism, he claimed:

> An astrologer will search in vain in my horoscope for the reasons why in 1595 I discovered the relationship between the heavenly spheres ... ⁴

Should one disagree with him on this matter? I asked a couple of astrologers as to whether they could see anything special about that inspirational afternoon, but they could not seem to discern much of note about it. To be sure there was a Mercury-Neptune conjunction that week, in trine to Pluto and Uranus, and Saturn was crossing his ascendant, but that did not seem much to account for that singular moment. Then, Mike O'Neill (who will be referred to hereafter as

Chapter 2. Quintiles, Septiles and Genius

M.O.) noticed that several of the little-appreciated septile aspects were present. Were they perhaps the key?

What Kepler might possibly have noticed (but did not) about that day, July 19th, 1595, was that there were three septile aspects in the sky. He was later to argue forcefully against septiles, and in doing this he was probably the only astrologer of his day who even considered them! On a chance basis, one would expect just one such septile aspect to be present. If we include the outer planets, then four septiles were present on that day, but the chance-expectancy level becomes just over two septiles.

The aspects formed by dividing the zodiac into five parts are called quintiles. As the theory which then came to him was somewhat fivefold in nature, let us note that there were no quintiles in the sky on that day — the cosmos was not supporting Kepler's fivefold theory of planetary distance! Later on, we will look at the patterns in the heavens over his illumination twenty-three years later, when the immortal third law finally dawned upon him after months of racking his brain over the figures —

> ...nothing holds me back.
> I can give myself up to sacred frenzy ...

— when there were also four septiles present in the sky. These were, in fact, rather stronger than those present at this early moment, for a reason to be discussed later: they had closer 'orbs', i.e. they were more exact.

In 1985, I began collecting nativities (birth data) of eminent scientists, together with the moments when they experienced their decisive breakthroughs, without too much idea of what to do with them. I had taken a degree in the History of Science years ago, and wanted to try and make some use of it. There was no notion as to what might be expected from these times, except that there should be some relation between the birth chart and the Eureka moments. If there were any truth in astrology, then there should be some link between them. As Kepler put it:

> And since almost every motion of the body or soul or its transition to a new state occurs at a moment when the figure of the heavens corresponds to the birth figure

(which is usually a matter only of certain correspondences in detail), it happens that some notable men will be most greatly moved by these Aspects, and others like them, since so many such men were also born under such Aspects.[5]

That is a statement about *transits:* when the planetary positions around the zodiac echo those of the natal chart of the person concerned, then that person will be 'most greatly moved.' The persons concerned in Kepler's example were princes of the realm, not ordinary people.[6]

John Addey's Hypothesis

The suggestion arose that 'quintile' angles were somehow linked to creative mental ability. This seemed in tune with a suggestion made by the late John Addey. His opus *Harmonics in Astrology* published in 1976 gave a new emphasis to these smaller aspects. The main theme of this book was that at the core of astrology lay what he called 'the harmonics of cosmic periods'. He argued that some kind of wave-phenomena might underlie the subject matter of astrology, which thesis has not as yet convinced many. We took from his book the above-quoted key statement, as to how the fifth and seventh 'series of aspects' figured in the charts of 'creative people.' Figure 2.2 shows the quintile and biquintile aspects as they are called (72° and 144°) and the different septile aspects, comparing them with a trine aspect (120°).

Table 2.1: Septile, Quintile & Trine Aspects using Addey orbs

Septiles	Quintiles	Trines
±1.7°	±2.2°	±4°
21° total span	19° total span	16° total span

On a first visit to M.O., a small computer on his desk was making a faint humming noise, which he declared was 'generating his control group', of 10,000 — or was it 100,000? — cases for an astrological hypothesis he was testing. The device seemed hardly larger than a pocket calculator, and it was a mystery to me how it could clock up

Chapter 2. Quintiles, Septiles and Genius

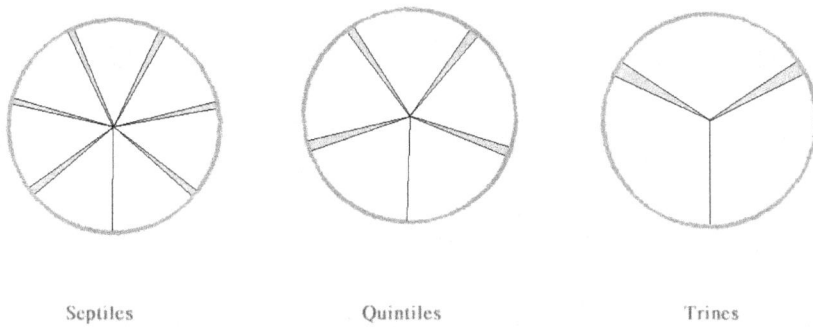

Figure 2.2: Quintiles, septiles and trine aspects

such a huge store of random data. One wire was plugged into his television set as its screen and another into a small tape-recorder, which was, he explained, its 'memory'. For the survey he was about to publish,[7] the computation of expected frequencies of the planets by a suitable random sampling procedure was a major challenge. As it was doubtful whether anyone at that time had ever achieved a positive result for a hypothesis concerning planetary aspects, the problem seemed quite academic.

M.O. suggested that we test John Addey's hypothesis on the data. His computer whirred back through the centuries at the touch of a few buttons, setting up the chart corresponding to a given moment of time. By four o'clock the next morning the hypothesis had been verified, to a fair level of statistical significance.[8] It was indeed the case that the eureka scientists had a net excess of these aspects. A moment we did not then manage to locate was Galileo's first sighting of the moons of Jupiter: Jupiter seemed to be below the horizon at the required hour. Eventually we found the correct time from an astronomical treatise, which had reconstructed the exact time of Galileo's observations from his diagrams of the moons of Jupiter: it was a matter of when Galileo considered his day as beginning.[9] The eureka project developed as a collaboration: I would believe M.O.'s computer programs, if he would accept my selection of eureka moments.

The prime-number aspects that we were scoring are readily obtained for a chart using an astrology program. But also, we had to find out how many of these aspects were expected by chance. Here M.O.'s expertise came into play, with his canny grasp of the ways

23

of computing expected frequency, shown in his earlier aspect study, referred to. I've put all these calculations into an appendix (appendix B), having reluctantly to accept that not many people have an interest in such computations. M.O. explained to me how the frequencies can either be calculated from probability theory, or empirically by sampling huge numbers of random charts, as his computer could do. For now, let's merely note that there is a fairly precise expected value, which lies between two and three, in the number of quintile and septile aspects normally expected per chart, as will depend upon orb.

An orb is the greatest angular distance in celestial longitude between two heavenly bodies, over which an aspect is believed to be able to exert some sort of effect. Or rather, it is the greatest angular deviation from a given aspect, such that that aspect will 'count'. Thus the above diagram (table 2.1) shows the orbs for quintiles, septiles and trines. Roughly speaking, astrologers are likely to use about two degrees of orb for quintiles, and perhaps one degree for septiles if they ever use them. John Addey gave a handy protocol for finding the orb of small angles, as follows. One takes a given span, which he suggested should be between twelve and fifteen degrees, and divides this by the *harmonic number*. Suppose we take twelve degrees, and that means plus or minus twelve degrees, as the basic span, then that would be his orb for a conjunction, i.e. a meeting between two planets in the sky. That's a rather outrageously large span, but hang on ... For a trine, which is a one-third division of the circle, the orb would be 4°, because one divides 12° by three.* Astrologers might find that

---

*Addey proposed that such orbs could be extended down to the several hundredth harmonic, so long as they were diminished in proportion; as if he had not seen the dire problem which thereby arose: using his orbs, the three trine aspects covered a total of 24° of the ecliptic, then the four square aspects another 24°, and each successive series of aspects (or rather, 'harmonic') covered a further 24° or so of the zodiac circle, so that the entire circle soon became covered many times. Every possible angle would thereby signify an indefinite number of aspects.

We here avoid this error, by accepting the limit which Kepler placed upon the number of effective aspects, or something resembling it. From his geometrical reasonings, he inferred that the lowest grade of configurations effective as aspects were the decile [36°], tridecile [108°], octile [45°] and trioctile [135°], (*Harmonice Mundi* 1997, Book IV Proposition XIV). Then he went a step further, however, adding that the smallest effective aspects were those limited by 'configurations which hesitate between power and powerlessness, namely the 24° arc from the

Chapter 2. Quintiles, Septiles and Genius

a little restrictive for a trine. The principle has been well expressed by David Cochrane:

> Note that in harmonic astrology, orbs must be proportionate to the harmonic number in order for the mathematical simplicity and elegance of harmonic astrology to survive.[10]

Cochrane chose to use larger orbs for his work by taking sixteen degrees rather then twelve as his base orb. Here we derive several 'harmonic' families using a twelve-degree orb: there are the two trine positions, then for the four possible quintile and biquintile aspects one divides twelve degrees by five, giving an orb of just over two degrees, and likewise dividing by seven gives the 'Addey orb' for a septile, of just under two degrees.

Using these orbs, two planets moving around the ecliptic will tend to be making one of the septile aspects for six percent of the time, and slightly less for the quintiles and trines. These small orbs mean that one is relying upon the computer producing spot-on positions for centuries gone by. In the 1980s, computer astrology programs were struggling to acquire such accuracy (say, to the arc-minute), and our computations all had to be re-checked using more accurate programs. Tiny Pluto, for example, wobbles about over the centuries on the gravity-fields of Uranus and Neptune, and quite powerful programs are needed to track it back to some earlier position.

Untimed birth charts were no use, mainly because of how far the Moon moves against the stars each day, and we needed well-defined planetary positions.† The situation is different for untimed E-moments, where, especially prior to the invention of the electric light,

---

pentakaidecagon [fifteen-sided] and the 18° arc from the icosagon [twenty-sided]' (Ibid, Propn. XV, p.245; quoted in Field 1984, p.218). In other words, any division of the circle into more than at the very most 20, i.e. smaller than 18°, is to be regarded as powerless. We here follow that advice.

†Birth data is usually given to the nearest hour; for comparison, the Moon moves half a degree per hour and one zodiac sign rises over the horizon every two hours. This means that what are called the 'angles' of a chart had to be omitted — i.e. the position of the horizon (called, the 'Ascendant') and of the MC (the highest point the Sun or any planet reaches in the day, the Medium Coeli or middle of the sky). For the small angles we are dealing with, one would need birth data accurate to within minutes to include these. The researches of the Gauquelins concerned this diurnal framework of the horoscope, i.e. the 'angles'.

there is a smaller uncertainty in the time-definition, say within four hours of noon for a daytime event, which turns out to be acceptable for our study. It corresponds to an uncertainty in the Moon's position of two degrees, or an average error of one degree, which is just about okay. So, we discarded the eureka scientists of unknown birth time. Sadly, this excluded a majority of them. Amongst the just over one thousand scientists included in Asimov's *BEST*, not much more than fifty have known birth times.

That is a tragedy, an irretrievable loss. John Aubrey's brave research project has not been continued. Aubrey started collecting birth data of contemporary characters in the seventeenth-century, immortalising them in his *Brief Lives*, sketching out their temperament, appearance and achievements *for the specific purpose* of furthering the investigation of astrology.[11] Few today would guess, from a modern edition of Aubrey's *Brief Lives*, that it was composed with an eye to investigating the truth of astrology by a collection of nativities. Alas, no-one asked Michael Faraday's mother in the East End of London as to when in the day her illustrious son was born, or perhaps she could not remember amidst her ten children.

The collection of eminent scientists' birth data assembled here would hardly have been feasible prior to the 1980's. In 1983, the Gauquelins published their large volumes of natal data Europe-wide for their eminent professionals — excluding England where the time of birth is still not recorded on birth certificates. Clearly a modern scientist would not wish to record so trivial and unimportant a piece of data. In America, Lois Rodden has assembled a distinguished nativity collection, graded according to its degree of reliability. In the UK, the Astrological Association's data collection is now very comprehensive, and has been a vital aid for the present study. Such a collection of nativities is a recent phenomenon, distinguishing the reliable birth data from that which is merely speculative.

Speculative birth data includes, for example, the claim that Isaac Newton was born at 1.00 a.m. on Christmas day, 1642. It is speculative because no-one knows a source for the time of 1.00 a.m. Data may not be speculative but 'dirty' if there are two different times quoted, as is the case for, say, Robert Oppenheimer. Lord Lister, founder of antiseptic surgery, is classified in Lois Rodden's 'dirty data' section because there are two different birth times, but as these times differ by merely one hour that was acceptable for us.

Our investigation was initially presented in 1986 as *Project Eureka* at the Fifth International Astrology Research Symposium at the Royal Free Hospital in London, and was then published two years later as *The Eureka Effect, an Initial Report*.[12] The tables and charts here used are essentially unchanged from that report, although there were one or two errors in the times used, and the data has been re-computed where necessary with these corrections inserted.

The beauty and inner coherence of our hypothesis lies in the way it applied *both* to the natal data of the E-scientists, *and* to their E-moments. That rescued it from the charge that the numbers in its groups are too small. John Addey's hypothesis was also applied to natal data, to the birth moments of creative scientists. We have extended it by applying it also to *conception* times for new scientific theories, viz. to the list of E-moments. We first tested our hypothesis upon the natal E-group, then moved onto the event group.

Scientists included had to feature in Asimov's *BEST*. Our study can only include the most eminent scientists who have biographies written about them. Isaac Asimov, in his potted biographies, has more or less always cited any E-moment where it was known, which was a help. However, one has to go to a full biography for an account of the E-moment. Many biographies were researched, and it was not felt necessary to cite them in this opus.

'My son, today I have made a discovery as important as that of Newton', Max Planck told his son Erwin Planck — but typically, neither he nor his son recalled the date. Indeed, it was only from Planck's *obituary* that the world learnt at all about this remarkable quotation! As his birth data is also unknown, he falls right out of our survey. If either of these were known, he would be included. We collected dated E-moments, plus undated ones for which the scientist concerned has known birth data. For Nicola Tesla, for example, the date of his great moment of inspiration in Belgrade Park is lost, yet that experience defines him as a eureka scientist. The loss of so many of the historic dates has weakened the conclusions which this survey is able to draw.

A New Element

> *What new element before us unborn in nature? Is there a new thing under the Sun? ... First penned unmindful by Doctor Seaborg with poisonous hand, named for Death's planet through the sea beyond Uranus.*
>
> Ginsberg, *Plutonian Ode*[13]

Why quintiles? Our appreciation of their significance as regards creative mental activity came about mainly from examining the creation of a new metal, plutonium, in relation to the chart of its inventor. The event has no direct relevance to this inquiry, as it involved no notable moment of realisation. However, perhaps by analogy or as an important occasion of birth, it has relevance. Let us turn to consider the creation of the unnatural element in Berkeley, California, by Glenn Seaborg.

I was looking for the date when plutonium was first made. Discussing the matter with astrologer Zach Matthews, formerly a civil engineer, then later editor of the UK's *Astrological Journal*, he emphasised that there was no use in looking at a chart unless one had obtained the definite, first-beginning moment, which would then be the birth time of the entity. That led to a letter being written to Professor Seaborg, gaining the time plus date for its first conception. It was this experience which formed my notion of 'genesis-moments', as *times of first-beginning*.

The letter from Seaborg had a copy of a page from his diary for the day when the process began, showing the time when the apparatus was switched on to within one minute. The Seaborg diaries have since been published,[14] but one might not have realised that this particular experiment was the decisive one which produced the first sample of the new element, without the letter indicating such. By another source, Seaborg's time of birth was also obtained. To have the birth time within an hour, and a genesis-moment to a minute, is a rare and valuable combination.

The new element was formed in the even-ing of December 14th, 1940 by Seaborg's team. Seaborg switched on a deuteron beam in the Berkeley cyclotron for four hours (8.00 pm until midnight), focused onto a sample of uranium. This created some 'neptunium', and over the next few days that formed the new element. This was a fine and terrible moment of birth.

What it showed, as M.O. pointed out, was a quintile in the plutonium chart between Pluto and Uranus, echoing the biquintile between the same pair in their natal chart of Seaborg — i.e. two-fifths of the circle or 144 degrees. These aspects were very close: within one degree and five minutes respectively. Quintiles were the new aspects discovered by Kepler, and many traditionalists object to using them, claiming that 'they don't work.' The quintile between Uranus and

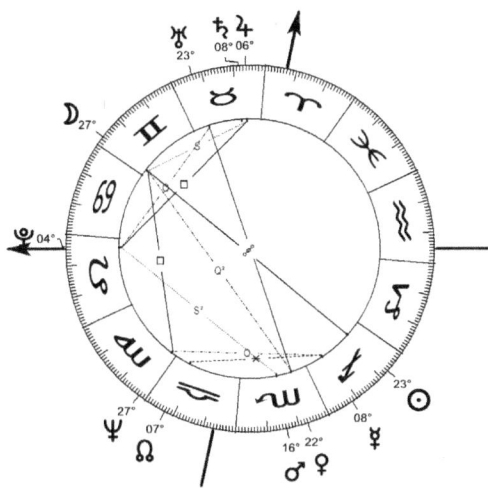

Figure 2.3: Event Chart for the creation of Plutonium — 14 December 1940 8 p.m., Berkeley cyclotron, CA

Pluto was strongly, indeed, hauntingly emphasised in the plutonium chart (figure 2.3): Pluto was just rising, conjoining the Ascendant to within half a degree, the Full Moon was exactly in the middle of that quintile in a square to Neptune, Seaborg's natal Pluto was exactly over that Moon, *to within five minutes*, and the whole axis of that Full Moon (bisecting the quintile) was lined up with the Galactic Centre at 26° Sagittarius.[15]

The emphasis on the number five in that chart expressed what was then making its appearance: plutonium has five valence states (these range from three — i.e. trivalent — to seven) which is extraordinary for a metal. It also has five crystalline conditions: to quote from a textbook which Seaborg wrote upon plutonium, this metal 'undergoes no less than five phase transitions between room temperature and its melting-point.'[16]

Thus, what he discovered seemed to have an accord with the celestial background present when it emerged.

The timing of this metallic genesis drew our attention to the quintile aspect as relevant for creative work. This aspect formed a clear link between the natal and event charts. We return to this topic in chapter 12. More recently, and for comparison, the authors of *Mundane Astrology*[17] have discussed the stages of unfolding of the quintile cycle, emphasising its link to creative mental activity, citing the example of Louis Pasteur. The biquintile was described by these authors as 'a period of very effective application of creative ideas and policies expressing its quintessential essence.' Pasteur was born on a conjunction between Uranus and Neptune, and reached his germ theory of disease in 1860–62, i.e. during the period of the quintile between these two.

References

1. John Addey, *Harmonics in Astrology* 1976, p.123.

2. Carl Sagan, *Cosmos*, 1980, p.57.

3. Kepler, *Mysteriumn Cosmographicum* 1595, 'The Secret of the Universe' English translation by A. M. Duncan, New York 1981.

4. Johannes Kepler, 'The Harmony of the World' (*Harmonice Mundi*, 1618) trans. Aiton et. al., Philadelphia 1997, p.377.

5. Published as 'Kepler's Belief in Astrology' by the author, in *Clio and Urania Confer: History and Astrology*, Ed. Kitson, 1989, pp.152–170.

6. Kepler, 'De Fundamentis Astrologiae Certioribus' 1602 Prague, para LXXV, translation in J. V. Field's 'A Lutheran Astrologer: Johannes Kepler', *Archive for History of Exact Sciences*, 1984 Vol. 31, p.268.

7. M. O'Neill, 'The Moon's Nodes in Synastry' *Astro-Psychological Problems* (Ed. F. Gauquelin, Paris), Sept. 1986, pp.24–30. For a less technical account, see the author's 'Darling, Where's your South Node?' *The Astrological Journal*, 1987, 5; summarised in 'Investigating Aspects' by N.K. in *Astrological Research Methods*, Ed. Pottenger, ISAR Los Angeles 1995 pp.287–302.

8. The work began on the evening of Friday, March 7th, 1986.

References

9. Jean Meeus, 'Galileo's First Records of Jupiter's Satellites' *Sky and Telescope* September 1962, 137–9. Galileo saw them on January 7th in 'the first hour of the night', i.e. an hour after sunset. However, it took him several days to figure out what was going on (for further discussion see p.56).

10. D. Cochrane, *Astrology in the 21st Century*, FL 2002, p.96.

11. Aubrey's *Brief Lives*, Penguin Classics, 1987; Patrick Curry, *Prophecy and Power, Astrology in Early Modern England*, 1989, p.73.

12. N. Kollerstrom and M. O'Neill, *The Eureka Effect, an Initial Report*, 1988. Strangely enough, a rather comparable 'Eureka Effect' had been described and so named at much the same time, by the late Brian Inglis: 'The Unknown Guest, The Mystery of Intuition', 1987, e.g. p.81. The Urania Trust reprinted *The Eureka Effect* in 1996.

13. Allen Ginsberg, *Plutonian Ode and Other Poems*, San Francisco, City Lights Books 1982.

14. *The Plutonium Story, The Journals of Professor Glen T. Seaborg* 1939–46, Ohio 1994, p.14. He started his plutonium-genesis on 14th December at 8.00 p.m. in the Berkeley cyclotron: 4.00 a.m. GMT, 15th Dec. 1940. Richard Rhodes, *The Making of the Atomic Bomb* 1986, p.353, unfortunately found a wrong date for this, around the beginning of January 1941.

15. 'Pluto and Plutonium' was published in the *Astrological Journal* of 1984, p.4.

16. J. Katz and G. Seaborg, *Chemistry of the Actinide Elements*, 1957, p.265.

17. N. Campion, C. Harvey and M. Baigent, *Mundane Astrology* 1984, p.160.

CHAPTER 3

The Act of Creation

'And, as imagination bodies forth
The forms of things unknown, ... '

   Theseus, Act V, scene i, *Midsummer Night's Dream.*

The View of Arthur Koestler

BORN IN BUDAPEST, ARTHUR KOESTLER became one of the best-known novelists and savants of his era. In 1964, he produced what he claimed was a contribution to psychology, *The Act of Creation*, though psychologists tended not to see it that way. This opus, together with another by a Russian-born author, will be one of the two main references for the present inquiry. That second opus is the *Asimov's Biographical Encyclopaedia of Science and Technology*, to be discussed later.

Figure 3.1: Arthur Koestler

Koestler's opus has been continuously reprinted since its publication, which is more than can be said for other books about psychology. The 'psychologists' of académe

33

may not have approved of it, but in the end did that matter? They have not been perceived to have much of note to say about the human psyche, let alone about the special kinds of moment with which we are here concerned.

During his schooldays, Koestler became convinced, as he wrote in *Arrow in the Blue*, that 'geometry, algebra and physics ... contained the clue to the mystery of existence', a mystery whose solution 'seemed the only purpose worth living for.' It may not be irrelevant to add, that he also became interested in the phenomena of psychical research. As a solitary youth, he was reading alone in a Viennese *pension*, when a tin of baked beans 'suddenly exploded and hit him on the head'. Was that tin of beans trying to tell him something? As the small detail of the date of that event is unrecorded, we are prevented from further comment.

Figure 3.2: The Eureka Moment — an alchemical view, M. Maier, Atalanta fugiens, Oppenheim, 1618. (*Wellcome Image Library*).

Later on, under sentence of death in a Spanish prison, Koestler apparently gained from mathematics some kind of illumination. He was whiling the days away working out the Euclidian proof that the number of primes is infinite, when suddenly he was 'filled with a certainty that a higher order of reality existed and that it alone invested existence with meaning.' Koestler became well-known for his criticisms of the reductionist slant of modern psychology; seen for example in his review in the *Sunday Times* of R. L. Gregory's *Mind in Science*, where he attacked professor Gregory's failure to discuss the 'nature of consciousness and its place in the physical world.'

The *Act of Creation* surveyed diverse experiences of illumination — from Gutenberg's brainwave about constructing a printing press in the fourteenth century, to Watson and Crick envisioning the helix of the molecule of life in the 1950s. Koestler was continually searching for that moment of altered awareness, perhaps a little similar to that

Chapter 3. The Act of Creation

which he had experienced once or twice, whereby the new paradigm could be realised. Logic took the mind along the familiar paths as patterned by tradition, but what could be said in words about those fairly rare illuminations which went beyond logic?

It is easy to dismiss Koestler's mode of describing these experiences as, say, too woolly or merely metaphorical, until one sees what others have done with them. The extreme inadequacy of more recent 'scientific' accounts to speak about these experiences, or what the eureka-type scientists have in common, drives one back to a renewed appreciation of what Koestler managed to do.

If anyone had noticed something special about Einstein's brain now pickled in formaldehyde, one would be keen to hear about it, but this seems not to be the case. Modern psychology rather lacks a vocabulary for describing the process of inspiration, though the notion of 'gestalt' and 'switch of gestalt' as developed by German psychologists is of relevance. The concept was employed by Kuhn in his epoch-making *Structure of Scientific Revolutions* to describe how scientific communities adjusted their perception of things in the wake of a 'scientific revolution'. As an example of a gestalt switch, Kuhn cited how the new planet Uranus was noted by several astronomers as a star, then seen by William Herschel as being a comet. Then finally a mathematician studying its motion declared that it was not a comet at all, but a new planet beyond Saturn. These shifts of perception by the scientific community are not quite what we are looking for.

The meaning which Arthur Koestler assigned to the proverbial cry, 'Eureka!' may be firmly anchored in the historical example of Archimedes. Using Koestler's terminology, we can say that two different matrices of logic came together at that moment. The king was suspicious that someone had been tampering with his gold so that the crown was a mere alloy, adulterated by a baser metal. The standard way of resolving the issue was by fire, remembered today in our word, 'proof'. A metal was put to the 'proof' by insertion into a furnace, so that the base metals and silver would melt and run away leaving the pure gold. But Archimedes would be in trouble if the royal crown was melted down. When getting into his bath his thoughts turned from fire to water, and the rest is history.*

---

*An appendix to Beveridge's *Seeds of Discovery* 1980 gives a translation of the

It is time to quote Koestler on the matter:

> ...new syntheses arise in the brains of original thinkers through the bisociation of previously unconnected matrices. The parallel process on the collective plane — on the map of history — is the confluence of two branches of science which had developed independently, and did not seem to have anything in common.' 'The progress of science', Bronowsky wrote, 'is the discovery at each step of a new order which gives unity to what had long seemed unlike.

What is first experienced by a single individual as a new illumination, then gains acceptance by the scientific community. Indeed, it is only a eureka experience if it has the result of changing the perception of the scientific community at large:

> The eureka act proper, the moment of truth experienced by the creative individual, is paralleled on the objective plane by the emergence, out of the scattered fragments, of a new synthesis ... [1]

We shall adopt this view, though it is unfashionable amongst today's science historians and psychologists, as assigning an unduly heroic role to 'the creative individual'

> ...that moment of altered awareness ... whereby the new paradigm could be realised.

Koestler designed The *Act of Creation* with a threefold structure: a first section considered humour, and how 'bisociation' as a process featured in jokes, by bringing in the unexpected, followed by a section on the history of science, and finally artistic inspiration. While admiring this grandiose project, its wider scope will not detain us,

---

account by Vitruvius (a Roman architect of the first century AD). This explained how 'information was laid that gold had been withdrawn from the crown,' while the same amount of silver had been added in its making. Gold changes colour if more than one-fifth of silver is added, so the 'fraud' which Archimedes discovered could not have been very large. Vitruvius' account was written three centuries after Archimedes' discovery of c. 240 BC (*De Architectura*, Book IX Ch 3).

Chapter 3. The Act of Creation

except to observe that it excluded the traditional religious dimension of inspiration.

Thus, Plato viewed inspiration as being a 'divine madness', averring that:

> ...all good poets, epic as well as lyric, composed their beautiful poems not by art, but because they are inspired and possessed ... there is no invention in him until he has been inspired and is out of his senses and the mind is no longer in him.[2]

This antique sense of the term, 'inspiration' is more drastic than any here considered, for it left its recipients amused or bemused as mere passive vessels. It was linked to the number nine, the nine Muses on Parnassus. Here, by contrast, we are dealing with septiles which are fairly mild ...

Koestler described the strong emotional charge associated with scientific eureka moments, a matter of key importance for locating their occurrence in the historical record. When difficult matters were suddenly resolved, after days or months of tension when all progress seemed to be blocked, this was he said 'always accompanied by the sudden expansion and subsequent catharsis of the self-transcending emotions. I have called this the "earthing" of emotion. The scientist attains catharsis through the reduction of phenomena to their primary causes; a disturbing particular problem is mentally "earthed" into the universal order.'

Thus, when the 30-year-old Michael Faraday succeeded in one of his first electrical experiments, his brother-in-law George Barnard who was present recalled: 'I shall never forget the enthusiasm expressed in his face and the sparkling in his eyes.' That was on Christmas day, 1821, in the house which is now the Royal Institution in London. A primitive electric motor had just been constructed, Faraday's response to the announcement the previous year by the Dane Ørsted about how an electric current deflected a compass nearby. The gladness was because a connection between electricity and magnetism had been resolved. It may serve as an example of what Koestler called the 'self-transcending emotions'. We have no time-of-day information for any of Faraday's discoveries, nor his birth chart, which limits further comment.

Before it happened, Faraday did not know exactly what he was aiming at, as no-one had ever produced rotation from an electric current. For several months he had been working on the problem. At the side of his notebook pages by the relevant experiments Faraday wrote 'electromagnetism.' We will return later to this achievement, as the chart is quite interesting even though it lacks a time.

'There is a tremendous emotional experience (in scientific discovery) which I think is similar to what some people would normally describe as religious experience, a revelation,' professor Charles Townes is quoted as saying.[3] Townes had the idea for what became the laser, of synchronous oscillation of energy within a single crystal. Brief quotes like these, which are quite hard to come by, confirm Koestler's view of the emotions involved in the eureka experience.

The condition preceding breakthrough Koestler described as a scientific impasse, a 'blocked matrix'. At such times, when progress seemed impossible, the scientist or mathematicians involved could sink into anguish before the solution was discovered. In this 'period of incubation' the person involved would sometimes withdraw from the stress of trying to tackle the main problem, and just let it lie fallow for a while. After the great breakthrough would come the business of consolidation, often involving battles with existing theories which had to be rejected. The new paradigm would then require verifying and elaborating, as it percolated through to the laboratories and schools.

Astrologically, one might associate the period of revolutionary turmoil with the planet Uranus and the more stable period of consolidation with Saturn; but also with Mars, because the old theories have to be destroyed for new ones to develop. Science does not resemble art or literature, where an innovation can co-exist side by side with some older style: the new theory can only be true if the old is abandoned, and lies shattered by the wayside. For this reason, relatively few people are interested in the history of science, because in contrast to the history of art its old theories are no longer experienced as valid. Perhaps this is why Mars as well as Saturn were found by the Gauquelins to pertain to eminent men of science.[4]

We will use the account left in Koestler's book as a template for deciding which events in the history of science have:

the basic, bisociative pattern of the creative synthesis:

Chapter 3. The Act of Creation

the sudden interlocking of two previously unrelated skills, or matrices of thought.

Normally the inner experience of the scientist is removed from any published report. He behaves as if his own psyche was the last thing in the universe that anyone's gaze would want to rest upon. Only if a biographer has been on the scene, collecting reminiscences from persons concerned, may we hope for adequate information to pinpoint these special experiences in space and time. From more formal biographies, such as those given in the grandiose, 15-volume *Dictionary of Scientific Biography* composed by the leading world experts, one will not normally find even the mention of the experience having occurred. Let us cite some examples given in the Koestler opus, to help focus the matter.

'Fortune favours the prepared mind' was a saying of Louis Pasteur's, known for his establishment of the germ theory of disease.[†] In 1897, he established the subject of immunology, the prevention of diseases by inoculation, so that he is nowadays mainly remembered as an immunologist. Koestler described how he achieved this 'by putting two and two together,' where the first 'two' pertained to the technique of vaccination, and the second 'two' was the hitherto quite separate research into micro-organisms.

In the spring of 1879, Louis Pasteur demonstrated that rabies and other afflictions were caused by micro-organisms, studying chicken cholera. He prepared cultures of the bacillus, but for some reason had to leave them until the autumn. When the chickens were injected with this aged bacillus, they became slightly ill but soon recovered. Using a new, virulent strain of cholera, some chickens soon died. But, to his astonishment, the chickens which had earlier been treated with the ineffective culture, all survived. An eye-witness in the lab described the scene which took place when Pasteur was informed of this curious development. Remaining silent for a minute, he then exclaimed as if he had seen a vision, 'don't you see that these animals have been *vaccinated!*'[5]

Up until that moment, the term vaccination only referred to smallpox, the vaccine which Edward Jenner had discovered. To quote

---

[†]It sounds better in French: 'le hasard ne favourise que les espirits préparés.' He was discussing Ørsted's discovery of electromagnetism, the previous year.

Koestler, 'The vision which Pasteur had seen at that historic moment was, once again, the discovery of a hidden analogy' — between his surviving chicks, and the bovine smallpox vaccine. Vaccination as invented by Jenner had been a common practice for three-quarters of a century, in both Europe and America, and it is hard for us nowadays to see why no-one thought of its wider significance earlier. After the great breakthrough, the new perspective becomes commonsense, while beforehand it was unthinkable.

Figure 3.3: Louis Pasteur (*Mary Evans Picture Library*)

Pasteur may have had an earlier eureka moment, when as a young man of 26 he discovered crystal optic rotation in a Paris hospital in May 1884, though recent examination of his notebooks has cast doubt upon this one. These are the two finest eureka moments produced in France, both lost.‡ There are some notable French invention moments, such as the first balloon ascending, but these are not quite so precious as are the E-moments.

The oldest documented eureka experience is Gutenberg's account of the invention of the printing press, in Mainz in the fifteenth century. A series of letters by Gutenberg describes his endeavour. He knew that cards could be printed using carved blocks of wood, with ink applied to their engraved surface: 'Well, what has been done for a few words, for a few lines, I must succeed in doing for large pages of writing ... What am I to do? I do not know: but I know what I want to do: I wish to manifold the Bible ... ' For months Gutenberg pondered such things as how seals were made of

---

‡Eureka experiences cluster predominantly around Britain, Germany and America. French scientists occasionally have dream-visions, such as Pascal having a vision telling him to give up mathematics for religion, or Descartes having the fabric of Cartesian space revealed in a dream (reviewed in chapter 14), but these are somewhat different.

Chapter 3. The Act of Creation

metal, but made no progress.

The answer dawned upon him during the wine harvest:

> I took part in the wine harvest. I watched the wine flowing, and going back from the effect to the cause, I studied the power of this press which nothing can resist ...

That pressure, he realised in a marvellously Saturnine eureka moment, could press down a seal of lead onto paper and leave a print. '... A simple substitution which is a ray of light ... To work then! God has revealed to me the secret that I demanded of Him ... I have had a large quantity of lead brought to my house and that is the pen with which I shall write.'[6] The first Bible was ready to be printed. That may be our sole eureka moment in which thanks were given to the Deity. All dates are lost.

Koestler commented,

> The ray of light was the bisociation of wine-press and seal — which, added together, became the letter-press. The wine-press has been lifted out of its context, the mushy pulp, the flowing red liquid, the jolly revelry ... and connected with the stamping of vellum with a seal.

What Gutenberg then discovered was how to make an invention work. As an innovation it was technological, applied science. Our list of dated eureka moments, more than twice as many as Koestler described in *The Act of Creation*, happens (somewhat regrettably) not to include any such technological E-moments. Another example of such is Edison's insight into how to make a phonograph, while pondering the rhythm of a morse code tapper. A recent US publication has located the date of this fine E-moment, described in detail by Koestler[7] and so establishes Edison for us as a eureka-type scientist.

As electricity and magnetism drew together in the year 1820, they sparked off some celebrated moments of insight. Hans Ørsted in Copenhagen had noticed during a lecture that an electric current in a wire caused a compass nearby to move round, a phenomenon so evident nowadays that it is hard for us to appreciate how mysterious it then seemed. In July 1820, Ørsted published an account of this historic observation, which he had made three months earlier, 'around April.'[8]

Following this publication, on Koestler's version:

> The news created an immediate sensation in Paris, where Ampère's excitable brain gave off a spark bigger than any Leyden Jar: he realised in a single flash that if an electric current produced a magnetic field, as the reaction of the needle indicated, then *all magnetic fields* may be due to electric currents ...§

The date for this E-moment is sadly lost, as likewise is Ampère's birth data.

Figure 3.4: The Andromeda Galaxy

The finest of American e-moments was located a decade ago in 1997. During 'The Night the Universe Changed Forever'[9] a vision dawned upon the astronomer Edwin Hubble, or possibly while walking down the side of a mountain in dawn's early glow. He had spent a night in the world's most powerful telescope, that of Mount Palomar in California, while it was pointed at the Andromeda nebula, in Orion's belt. It had taken five years just to grind the lens of the Mount Palomar telescope, and that night it disclosed to Edwin Hubble a more immense universe than had been dreamed of hitherto. Returning to his office the next morning, Hubble checked over the photographic plates he had just taken, and realised that this had to be a galaxy in its own right, just like the Milky Way.

Hubble's biographer gave both the date and time of his E-moment, which is remarkable.

He described how Hubble was forced to a remarkable conclusion: that the Andromeda nebula was

> ...at least 300,000 parsecs from earth — the equivalent of a million light years, or more than triple the diameter of Shapley's entire universe [Hubble's mentor, Howard Shapley, had proposed that everything known was contained within the Milky Way]. It was a 'eureka' moment.

---

§Ørsted's discovery was announced to the Paris Academy of Sciences on 4 Sept 1820, followed on 25 Sept by Ampère's announcement.

Andromeda's spiral arm was bejewelled by a Cepheid variable. The giant nebula was a sister to the Milky Way, composed of stars by the millions! Hubble crossed out the letter 'N' for nova he had previously inscribed on the plate and printed 'VAR!' for variable directly beneath it. And though the plate had been taken the previous night, he dated it '6-Oct 1923' to commemorate the moment when his mental tumblers had fallen into place.

Alas, we are told nothing about how Hubble *felt* at this galactic moment of illumination, after gazing at the glittering majesty of that starry wheel in the depths of space, but that is often the way with scientists. He spent the next few months in a feverish search for evidence to confirm his new view. Whereas previously there had been only one galaxy, that of the Milky Way, Andromeda was now appearing as another. Hubble had re-envisaged what had earlier been merely a 'nebula', i.e. a patch of cloud.

That will suffice by way of illustrations concerning Koestler's general thesis. It is a theory about creative individuals at key points of history, not about historical determinism. Afterwards, in retrospect, it may look as if the insight was determined by what went before, but at the time the individual seems to usher in the future by an act which is their personal contribution to history.

When Paradigms Shatter

Some readers may wish to place the Koestler account of the eureka experience within the wider context of that provided in Kuhn's sociological study, *The Structure of Scientific Revolutions*, published in 1962. We should not criticise Koestler for failing to notice this seminal work, deeply connected with his own argument, a mere two years prior to his own. Our main reference works were published in the early 1960's![10] One can add in here the Asimov reference, published in 1964. There is one work of importance published in the 1970's to which we shall refer, however it is astrological.

The word 'paradigm' used above came from Thomas Kuhn, who effectively gave it the meaning it has since retained, namely the fabric of theory elaborated during periods of 'normal science'. Normal science he viewed as a fairly staid process whereby matters progressed on metal rails with members of the scientific community feeling secure because there was communal agreement as to what had been explained. Things here and there which no-one could explain were regarded as problems to work on, not challenges to the whole 'paradigm'. But eventually, one advanced into a period of crisis, when again no-one was sure about the basic axioms. As an example, Kuhn cited Wolfgang Pauli, who, some months before Heisenberg's paper on matrix mechanics pointed the way to a new quantum theory, wrote to a friend: 'At the moment physics is again terribly confused. In any case, it is too difficult for me, and I wish I had been a movie comedian or something of the sort, and had never heard of physics.' Only a few months later, he wrote 'Heisenberg's type of mechanics has again given me hope and joy in life. To be sure it does not supply the solution to the riddle, but I believe it is again possible to march forward.' Heisenberg in a fine eureka experience had resolved the problem which tormented Wolfgang Pauli.

During what Kuhn called 'extraordinary science' the eureka experiences were supposed to occur. Kuhn's critics doubted whether history could be dichotomised into normal versus extraordinary science, but his view of scientific revolutions has been most influential. 'Extraordinary' science was supposed to be a time of uncertainty, when the guiding principles become uncertain: 'Though there is still a paradigm, few practitioners prove to be entirely agreed about what it is.' At such times, Kuhn explained:

> ...the new paradigm, or a sufficient hint to permit later articulation, emerges all at once, sometimes in the middle of the night, in the mind of a man deeply immersed in crisis. What the nature of that final stage is — how an individual invents (or finds he has invented) a new way of giving order to data now all assembled — must here remain inscrutable and may be permanently so.[11]

One is impressed by such humility. The new paradigm may not emerge all at once. In scientific revolutions as Kuhn defined them, whether sudden illumination strikes the key person concerned or whether insights are reached gradually is immaterial. Kuhn's essay was on the sociology of the history of science and so not vitally concerned with this issue.

Of Youth and Age

Today's prevalent myth about the youthful nature of the revolutionary innovator was affirmed by Thomas Kuhn:

> Almost always the men who achieve these fundamental inventions of a new paradigm have been either very young or very new to the field whose paradigm they change ... obviously these are the men who, being little committed by prior practice to the traditional rules of normal science, are particularly likely to see that those rules no longer define a playable game ... This generalisation about the role of youth in fundamental scientific research is so common as to be a cliché. Furthermore, a glance at almost any list of fundamental contributions to scientific theory will provide impressionistic confirmation. Nevertheless, the generalisation badly needs systematic investigation.[12]

We can here perform such a systematic investigation, as our eureka study was based on the most comprehensive group of such moments ever collected. The claim turns out to be radically mistaken. The average age for the eureka experience falls in the late thirties. From 32 eureka experiences, of which 21 can be dated, the average age is thirty-seven. There is today a prevalent attitude that once over 30 one's creative mental life is virtually over, but history does not support such a view. Likewise, a catalogue of invention moments, when inventions first worked, gives a similar age distribution.

As for Kuhn's claim that 'almost always' the innovators should be, 'very young', or at least 'very new to the field whose paradigm they change', there is the occasional supportive example such as James Watson, whose chief expertise seems to have been in ornithology until a few years before he cracked the double helix, however this is a rare exception. Eureka breakthroughs can sometimes occur in the twenties, for example Charles Darwin arrived at his 'natural selection' theory when only 29, James Watson found the DNA pattern when 25, and Werner Heisenberg envisioned the 'magic matrices' of quantum mechanics when only 24. These are however exceptions. Normally, the persons who achieve the great innovations are those deeply enough immersed in their subject to perceive its necessity, and these are far from being young newcomers on the scene. The question as to how such an erroneous view came to seem quite self-evident to Kuhn may be worth raising.

In *Scientific Genius, a Psychology of Science*, D. K. Simonton attributes to Einstein the saying that: 'A person who has not made his great contribution to science before the age of thirty will never do so.' While it is unlikely that Einstein ever made such a remark,¶ suffice to say that Pasteur was 57 years old when his insight created the field of immunology, Ampère 45 when he grasped the link between electricity and magnetism in 1820, Galileo 46 when he discovered the moons of Jupiter, Kepler 46 when he discovered his third law of planetary motion, Mendeleef 35 when he discovered his Periodic Table, Alexander Fleming 47 when he discovered penicillin, Röntgen 50 when he discovered X-rays, and Lise Meitner 60 when she discovered the principle of nuclear fission. Our inquiry shows clearly that moments of creative illumination occur over the whole span of life, and the distribution has no skew towards the youthful twenties as the experts seem to believe.

Newton had his major creative period around 1685 when he hammered out the implications of the law of gravity, at 43 years of age. The popular image has it all coming to him in the year 1666 under his mother's apple tree, when he was merely 24, and as this image seems to occupy a vital niche in the national psyche no doubt it has contributed to the 'youthful genius' image of the scientific innovator. After three centuries, science historians are finally facing up to the shortage of documents which would establish that anything resembling a theory of gravity came to him at that youthful age. Much of his light and colour theory dawned upon him around that date, so did his early thoughts of calculus as 'fluxions', which was enough for him to be getting on with.

¶Einstein's biographer Abraham Pais knew of no source for such a remark, as made by Einstein.

References

1. Koestler, Arthur, *AOC*, 1964, 1989 Arkana, p.225.

2. Plato, *'Ion,'* Loeb Classical Library, Plato VIII 1925, pp.421,3 (thanks to Julia C.)

3. Berland, T., *The Scientific Life*, MIT 1962.

4. Gauquelin, M., *Written in the Stars, The Proven Link Between Astrology and Destiny*, 1988, pp.106-7. Regarding the validity of the Gauquelin endeavour, see the author's 'How Ertel Rescued the Gauquelin Effect' *Correlation*, 23, (1) 2005. http://www.astrozero.co.uk/astroscience/koll1ge.pdf.

5. *AOC*, p.112.

6. *AOC*, p.123.

7. *AOC*, p.197; For the date of Edison's E-moment, see: Hughes, T. *American Genesis* 1989 p.75.

8. R. Stauffer, Ørsted's Discovery of Electromagnetism' *Isis* 1957, 48, pp.33–50; also *Isis* 1953, 44, pp.307–310. Ørsted's lab. notes have been published (Skrifter, 1920, Copenhagen), so a date may be obtainable. A wrong date of 'Winter 1819' was given in some reports.

9. Gale Christianson, 'The Night the Universe Changed Forever', *The Griffith Observer*, June 1997, pp.4–10; Ibid, *Edwin Hubble, Mariner of the Nebulae*, Bristol 1995, p.158. Immanuel Kant had earlier envisaged this: Timothy Ferris, *Coming of Age in the Milky Way*, 1998, p.149.

10. Some further cases were discussed in Beveridge, *Seeds of Discovery 1980*; e.g., it dawned upon Charles Nicolle, Director of the Pasteur Institute in Tunis, that typhus was transmitted by lice, while stepping over a corpse one day on his way into the Institute, which Beveridge found to be 'an example of a eureka intuition'. Other Eureka-ish studies are: Augustine Brannigan's *The Social basis of scientific discoveries*, 1981 (sociological); Brian Inglis *'The Unknown Guest, The Mystery of Intuition'* 1987, with a chapter on 'The Eureka Effect' (paranormal emphasis); Karl Simonton's *Scientific Genius, a Psychology of Science* 1988 (a stultifying, behaviouristic approach), and R. Roberts, *Serendipity, Accidental Discoveries in Science* 1989 (insightful and chemically-oriented but lacking in dates); *The Eureka! Moment, 100 Key Scientific Discoveries of the 20th century* 2002 by Robert Lee (gives no dates).

11. Thomas Kuhn, The *Structure of Scientific Revolutions*, 1962, p.90.

12. Kuhn, Ibid, p.90. The issue is further discussed in Beveridge, ref. 10.

CHAPTER 4

The Angles of Inspiration

'...the Septenary is sure to appear whenever the human being traverses a hidden structure of time or space.'

Joscelyn Godwin, *Harmonies of Heaven and Earth*[1]

IN HIS PIONEERING WORK, *Harmonics in Astrology*, published in 1976, John Addey argued that the significance of a given aspect was in part derived from the meaning of the number associated with it — the number by which the circle of 360° has to be divided in order to arrive at that aspect.[2] Such an argument is rather Pythagorean, involving the notion that the number seven, say, has a significance, a quality, which is its own irrespective of any specific context.

That was a fairly new idea in astrological circles. Addey also claimed that aspects not normally used by astrologers, such as the septile and quintile series, were of great importance. While no history of these matters has as yet been written, it seems likely that prior to his work, astrologers were not in fact taking any notice of them. Persons one has asked on this matter, such as Sue Tompkins, author of a book on aspects, or Charles Harvey, co-author of a book which looked in some detail at the fifth and seventh 'harmonics' in astrology, were not aware of anyone having been in the habit of using these aspects prior to Addey's work.

To some extent the quintiles were in use, being amongst the new aspects introduced by Kepler. We find these described in for example an eighteenth-century astronomy (repeat, *astronomy*) textbook as: the quincunx of 150°, the quintile, of two signs and 12°, and the biquintile, $\frac{2}{5}$ parts of the whole circle, of four signs plus 24°.[3] The advent of the home computer program has made these aspects readily accessible, and is clearly the new influence which has brought these aspects into fashion — or, at least, in use amongst the avant-garde. A mere press of the button brings up the required 'harmonic'.

For readers not familiar with these issues, the traditional aspects were always formed between the same degrees of the zodiac signs, and so could easily be spotted in a chart. The numbers associated with them were all sub-multiples of the zodiac number 12, viz: one, two, three, four and six. Thus 27° Cancer and 27° Capricorn signify positions 'in opposition' i.e. on opposite sides of the zodiac. We have seen how it was Kepler's radical scepticism over the notion of the zodiac which enabled him to break out of its spell and be the first to define his three extra-zodiacal aspects. However, they do not exactly leap to the eye in a normal chart and astrologers as a group may not have possessed Kepler's keen mathematical discernment.

By the term 'series', as in 'the quintile series', we simply mean the quintile and biquintile, corresponding to 72° and twice that, 144°. In a sense there are four quintiles, but as they are symmetric about the 0° – 180° axis, there are effectively only two distinct types. Using the term 'harmonic', as in 'the fifth harmonic' to describe them, includes all of these four different quintile aspects, so that by using it we are assuming that there is no difference in quality between them. 'Harmonic' is a word having several differing meanings, and it could be unwise to fix it too definitely.

The idea of 'harmonic' aspects can be readily explained in terms that are computer-friendly. After all, a computer does the hard work in our investigation. If we consider the sixth harmonic, this will give five different angular positions between two planets. A one-sixth division of the circle gives a sextile, which is sixty degrees in span. Yet, there are only two sextile positions from any given point in the zodiac, so where have the other three come from? They comprise one opposition and two trines — but, at a smaller orb. These other aspects (whose harmonic numbers are two and three) are included, but at a smaller orb. That's the procedure we used. One might object on the grounds that trines have a different effect from sextiles, and shouldn't be lumped together in this way. That dilemma need not perturb us unduly, however, as we are here mainly concerned with prime-number harmonics.

One could include the conjunction as well, as would give six different positions to the sixth harmonic, which is simple enough. Addey actually proposed this in his *Harmonics in Astrology*, but we preferred not to do that. Addey had in mind the image of a regular waveform, on which model the seventh harmonic would have seven peaks, and so forth. Inclusion of the conjunction makes things very much more straightforward for the computer, because it makes the periodicity for each pair of planets in the heavens quite regular, and so irons out the expected frequency of a given aspect over epochs of time. That may suffice for now to indicate the choices available in producing a definition of 'harmonic' suitable for computer work.

Three centuries before Addey, Kepler's equally Pythagorean views on the significance of celestial angles had taken a different tack. He pondered the musical concordance of the ratios which they embodied, and showed how regular star-polyhedra were related to them, and so deduced which were effective and which inoperative, from notions of symmetry and congruence (to use a geometrical term). Addey's approach was more arithmetic, of claiming that numbers had an inner significance, from which one could infer root meanings of astrological aspects, and that was something new. Kepler claimed that the septile angle was inoperative, because the musical ratios corresponding to it such as 6:1 were discordant, but Addey reached no such view. Addey's scheme has the disadvantage of generating too many angular relationships which it claims are meaningful — while Kepler kept the number of potent aspects on a tight leash as it were, limiting their proliferation by his stern geometric reasonings.

The Characteristics of the Quintile

...the irregular and vital beauty of the pentagram

Claude Bragnon, US architect (1866–1946)

It appeared to Johann Kepler, from contemplating the musical harmonies underlying the theory of astrological aspects, and also from reading Ptolemy's unfinished work, *Harmonia*, that the ratios 1:4 and 2:3 ought to be significant for astrologers, in addition to the traditional ones they had been using since time immemorial. These were obtained by dividing the circle into 5, and he named them the quintile and biquintile, thus:

Table 4.1: 5th harmonic aspects

	Angle	Ratio
Quintile	72°	1:4
Biquintile	144°	2:3

In recent decades, astrologers have come to take seriously Kepler's new angles, linking them to mental qualities: to creative activity and to one's art in life, using that word in a wide sense. In *Harmonics in Astrology*, John Addey claimed that the number five represented: 'the putting together of form and matter and in this sense art,' and added: 'Notice that this process is accomplished by mind, for it is mind which can subjectively take into itself the idea of formal principle and the idea of matter and so unite them. Hence, note, a relationship between Five and *mind*.'[4]

I like the recent evaluation of quintiles by US author David Cochrane:

> The 5th harmonic has an innocent enjoyment in the play of life. There is an intelligence in the 5th harmonic ability to play with ideas. Practicality and seriousness are so unimaginative and deadening to the creative spirit of the 5th harmonic. The 5th harmonic manifests in a great number of ways from a love of culture, literature, or movies to a love of games and puzzles, in disinterest in the responsibilities and worries that seem to dominate other people's lives. Because the 5th harmonic is not concerned with practical applications, it is free to be imaginative, and sometimes more intelligent than those who ground themselves in applying, disciplining and controlling their talents in more respectable fashion. When a person with a strong 5th harmonic becomes passionate, often that passion concerns a book, movie, game, art, or

Chapter 4. The Angles of Inspiration

personal creativity that may seem relatively unimportant to others ...

The essential energy process of the 5th harmonic is extending oneself beyond the domain of mundane and practical affairs into the realm of play, creativity, or culture.[5]

A somewhat erroneous view was expressed by David Hamblin in his *Harmonic Charts*, where he said: 'My own research has convinced me that Fiveness is essentially connected with the idea of *making, arranging, building, constructing, structuring, forming.*'[6] It is rather hard to see how the number five can be connected with a principle of building: the honey bee builds with the hexagon and man builds with squares, but what can be built with a pentagon? In this work, we tend to perceive the prime numbers five and seven as having a mould-breaking character, somewhat by way of contrast to the more symmetric and even numbers two, four, six and eight, which are used in building a firm and stable structure.

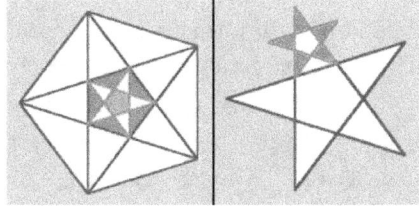

Figure 4.1: Pentagons and pentagrams

Figure 4.2: Pentagrams and hidden pentagons

Concerning the association with mental activity cited for this number, it may be of interest to look at how Isaac Newton used it, as has been quite sensitively depicted by Castillejo, from his study of both published and unpublished works. Castillejo found that number symbolisms were used both in Newton's published work and in his private, unpublished manuscripts (theological/alchemical), and affirmed that the symbolism was:

... most noticeable in the fifth, which always describes the attack or moment of contact or invasion, both in his scientific and prophetic works ... Then in the *Principia* it is the fifth Definition that he first mentions the attractive force of gravity ... And in his Optics it is at the fifth Axiom that he gives his first diagram showing the entry of a ray of light into ... water.[7]

This somewhat tends to confirm our remarks about the association of the quintiles with mental activity.

The five-sided pentacle or pentagram has the remarkable aesthetic property that its sides intersect to give the Golden Ratio, or 'divine proportion'. Mathematically, this ratio is expressed in terms of the number five. From the pentagram one can form a special triangle with two quintile angles at its base, and whose sides form the Golden Ratio.*

For centuries it was believed that the pentagram had a fiendish power whereby inscribing it upon the ground might provoke something unexpected (and dire) to emerge from it. This is not unrelated to our associating the fivefold pattern with creativity. Something unexpected may happen, and one cannot quite say what the devil is going on.

The pentagram unfolds through living things, from flowers to sea-creatures, but is not found in crystals. It is a number which flourishes wherever things bloom and burgeon, via its link with the golden ratio —

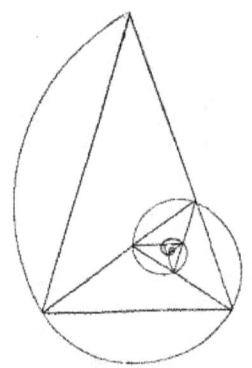

Figure 4.3: Golden Section/Spiral

and thereby to the Fibonacci series. As US numerologist Michael Schneider expressed the matter: 'Pentagonal symmetry has long been revered due to its profound insight into living nature and to the powerful psychological hold it has upon people throughout the world. It manifests itself in surprising ways and places in art, crafts, archi-

---

*The golden ratio, defined by $x : y = y : x + y$, is equal to half of one plus the square root of five. I like the idea that the cosine of a quintile equals half of the golden ratio: J. Kapraff, 'Connections, the Geometric Bridge between Art and Science', 1991 p.82; Matila Ghyka, 1977.

tecture, religion, magic ritual, national icons, and much else that is rooted deep within us.'⁸

Significance of the Septile

There is something hidden about the number seven, for it is not found in Nature. The passion flower has ten petals, the lily has six, the apple blossom five, the iris three parts, the cat has four legs, a crystal eight corners, and conception takes nine moons, but what has seven? To quote Kepler:

> We cannot find any body or other thing of this world that was made and specified by God on the basis of a septangle, nonangle [$\frac{1}{9}$ of a circle] or a undecangle [$\frac{1}{11}$] of a circle. Therefore it is also the case that nature does not enjoy any proportion that would be derived from such rejected figures.⁹

We cannot wholly agree with Kepler on this matter, because nine is not absent from nature in the same way: there are just nine lunar months on average between conception and birth. Nine are the planets which circle the Sun, eleven are the years of the solar cycle, and so forth.

Locusts in the desert breed to 13- and 17-year cycles, having adopted such prime numbers to avoid the simpler periods used by their predators. Thus one should not suppose that the prime numbers from seven upwards were merely not required by the Creator, Mother Nature

Figure 4.4: Giza pyramids: septile angle

or whatever. It is only the number seven that appears hidden from the realm of nature. Only in the vision of the totality of things can it be found, as in the Periodic Table of the chemical elements, which is a sevenfold scheme.

It would thus seem that the number seven is rather hidden and inexpressible, even opaque to the rational mind. Traditionally, it was a number with deep and somewhat religious significances, as the

Seven Pillars of Wisdom and the seven Days of Creation in the Old Testament, from which (traditionally) there followed the seven days of the week. The slope angle of the Great Pyramid is a one-seventh division of the circle. Through the millennia people have gazed at this angle, which is a septile to within about half a degree.[†] This helps us to appreciate the mysterious and unfathomable nature of the septile.

Kepler adopted a rational approach to the mysteries of the heavens, and so not surprisingly arrived at a dim view of the septile angle. This is the one issue on which we are obliged to part company with him:

> As God the Creator played music, in like manner He taught Nature to play after His likeness ... that is to say, precisely that piece of music that he has played for her. Therefore it is the case that in music no natural soul of a human being wants to play with a septangle, nor will he enjoy such if this proportion is given to the voices, because God did not play with these figures.[10]

Did He not?

Well, there *is* one manner in which this hidden number does in a sense appear in nature: namely, that a maximum of seven eclipses can occur in one year. One would have to jet about the world to see them. Surely, rather than contradicting what we have said above, this rather confirms the cosmic and slightly awesome and religious aspect of this number.

Esoteric meanings of the number seven, and its relation to scales of music, are discussed in *Harmonies of Heaven and Earth* by Joscelyn Godwin,[11] though such matters may have little relevance for the present study. Modern physics employs seven fundamental units, no more and no less. The Systeme Internationale or S.I. system of

---

[†]The three Giza pyramids (figure 4.4) have quite similar slope angles, whose mean is $51°\,40'$, the one-seventh angle, within a quarter of a degree. The slope angle of the Great Pyramid is given by $\arctan(4/\pi)$, i.e. its base perimeter is equal to the circumference of a circle whose radius equals its height, giving a slope angle $\alpha = 51°\,51'$; but also, marvellously, by $\arctan(\sqrt{\Phi})$ where Φ is the golden ratio, from the condition that the area of a pyramid side equals the area of a square on its height, giving $\alpha = 51°\,50'$. Its slope angle is between these values, while the septile is $51°\,26'$; Ghyka, 1977, p.22, 67.

Chapter 4. The Angles of Inspiration

units, is the system nowadays agreed upon worldwide by physicists. Its seven basic units, in terms of which all other units are defined, are: the metre, the kilogram, the second, the ampere, the Kelvin, the candela and the mole.[12]

Isaac Newton used the number seven to signify completion and wholeness: his optical theory of 1675 was immortalised by its seven colours, his little-appreciated lunar theory of 1702 comprised seven stages, his book *Optics* was composed in seven sections,[13] and his 'Newton's rings' diagrams had patterns of seven rings. All his alchemical notes had seven metals — as indeed did everyone else's at the time. On the other hand, in his unpublished works the number seven is given a more traditional mystical significance, deriving from his detailed study of the Book of Revelation and other prophetic works. His account of Solomon's Temple design, written in his last years, rejoiced in seven-branched candlesticks, the sound of seven trumpets, the pouring out of seven vials, seven gates and seven keys.[14]

When Newton announced that there were seven colours in a spectrum, inserting a seventh that no-one could see, indigo, his ruse was immensely successful owing to the traditional associations of that number, as signifying completion. In his day there were seven 'planets', meaning seven spheres which visibly moved across the sky, as there were seven metals: gold, silver, mercury, copper, iron, tin and lead. No-one then viewed the semi-metals arsenic and bismuth as real metals. The apparent correspondence of seven metals with the seven planets over nearly two millennia imbued the number seven with a very firm meaning which we have rather lost. Newton's letter proposing the seven colours of the spectrum played on the analogy with the seven notes of the musical scale, emphasising the sense of wholeness associated with this number.

If we take the first ten integers and divide the circle by them, the angles we obtain are as follows:

$360° \div 2 = 180°$
$360° \div 3 = 120°$
$360° \div 4 = 90°$
$360° \div 5 = 72°$
$360° \div 6 = 60°$
$360° \div 7 = 51.428571428571428571428571428571\ldots°$
$360° \div 8 = 45°$
$360° \div 9 = 40°$
$360° \div 10 = 36°$

There would appear to be something inexpressible about the number seven. If we divide seven into unity, it gives 0.142857 recurring, a figure which has strange sequential and cyclic properties ideal for a winter evening. Multiplied by any number from 1 to 6, it gives a number that contains the same figures, but in different sequences. Multiplied by 7, it gives 0.999999, while split into 142 and 857, these added together give 999. Multiplying by 360 gives, as can be seen, the same sequence.

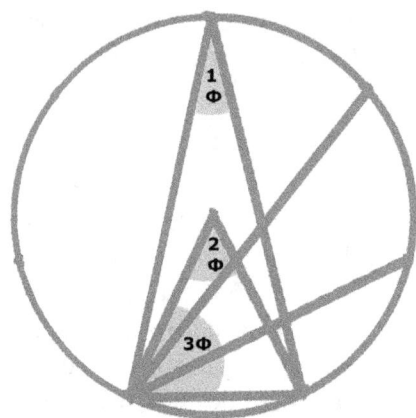

Figure 4.5: Dividing a circle into seven

The heptagon, that is the regular polygon with seven sides, is mathematically speaking not constructible. By this we mean, that the heptagon cannot be drawn using a ruler and compass only. The other polygons up to that number can all be constructed. Starting with the triangle, regular figures with each number of sides can be drawn. Mathematicians were finally able to stop worrying over this matter in 1801, when Gauss proved that the heptagon was not constructible (it followed from his proof that one could not trisect an angle, see figure 4.5).

There are three types of heptagon (figure 4.6), just as there are three septile angles. Starting from the centre of a circle, dividing it by one-seventh intervals i.e. *mono-septiles* will make a regular heptagon. Next, we move round that centre at two-seventh intervals, going twice round the circle and this gives us the 'alternate heptagon.' This is the *biseptile* angle of 103° to the nearest degree. Lastly, moving round

Chapter 4. The Angles of Inspiration

the circle in *tri-septile* angles (154°) and rotating three times, gives us the glorious star-heptagon.

We here tend to view these three types of septile angle as similar in effect, grouping them all together as the 'seventh harmonic.' In her book, *Aspects in Astrology*, Sue Tompkins omitted the septiles. As an aspect, it may somewhat lack a mundane meaning. In the view of John Addey, the septile angle was connected:

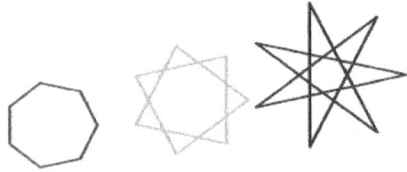

Figure 4.6: The 'family' of heptagons

> with sacred matters, with one's creations and creativity and with inspiration and one's receptivity thereto. To this we may add that it is evidently connected with the unitive and mystical aspect of things and with wholeness and so with the idea of fulfilment and completion.

An example Addey gave was the close Sun-Mars septile of Sir Winston Churchill, which he found to be the strongest solar aspect in Churchill's chart. He found this septile relevant to the imparting of dynamic inspiration to the Allied cause, this being Churchill's 'special genius' as was brought out by war.‡ Solar aspects for one so connected with the wellbeing of Britain are clearly important. Most of Addey's examples were drawn from artists, and neither Addey nor others discussing these aspects have given them any scientific context — at least not prior to the publication of the initial *Eureka Effect* report in 1988.

Astrologer David Hamblin endeavoured to develop Addey's views concerning these harmonic-aspects, claiming that seven was 'a number not of man's rational and constructive abilities, but of his wild, fertile and unpredictable imagination. It is a number of inspiration ... It is a number, not of permanence, but of transience; not of unchanging truth, but of sudden flashes of light and darkness; not of objective reality, but of subjective impressions and the emotional

---

‡Addey omitted to state that Churchill's Sun was immediately conjunct Antares ($\frac{1}{2}$°), the 'Heart of the Scorpion', thereby exaggerating the significance of his Sun-Mars septile (*Harmonics* p.123).

experience which man derives from those impressions; not of knowledge, but of fantasy,' concluding: 'Seven seems to relate more to man's fantasies about the universe than to its objective reality'.[15] That is hardly a view which will here find support.

Summing up, Addey found that seven was 'the number which represented the idea of creation,' whereas five 'may be said to represent the urge to power which is the prelude to creation.' He explained that:

> The question of whether 5 or 7 should be regarded as the number of creativity amounts to this: do we say that a man is creative if he has lots of good ideas as to how things should be done or made (2 + 3), or do we reserve this adjective for the man who not only has the good idea but actually puts it into practice and so produces something (3 + 4)? ... Seven also represents the influx of inspiration.[16]

References

1. Joscelyn Godwin, *Harmonies of Heaven and Earth* 1988, p.137.

2. Sue Tompkins, *Aspects in Astrology* 1989, p.41.

3. Charles Leadbetter, *The Compleat Astronomer*, 1742, Vol. 1.

4. John Addey, *Harmonics in Astrology*, 1976, pp.111–112.

5. David Cochrane, President of the (US) International Society for Astrological Research, *Astrology for the 21st Century*, 2002, p.85.

6. David Hamblin, *Harmonic Charts*, 1983, p.48.

7. David Castillejo, *The Expanding force in Newton's Cosmos*, Madrid 1981, p.103.

8. Michael Schneider, *A Beginner's Guide to constructing the Universe, The Mathematical Archetypes of Nature, Art and Science* 1995 p.97.

9. Ken Negus, *Kepler's Astrology*, Excerpts, selected, translated and edited by Prof. Negus, Princeton, New Jersey, 1987, p.10; quotation from Kepler's *Harmonice Mundi* of 1619.

10. Negus (ref. 10), p.13.

11. Godwin, (ref. 1), pp.137–8.

References

12. Keith Gibbs, *Advanced Physics*, Cambridge 1988, p.4.

13. Castillejo, (ref. 7), p.32.

14. Ibid, p.103.

15. Hamblin, (ref. 6), pp.64–5. Around 1990, David Hamblin announced that he had lost his belief in astrology.

16. Addey (ref. 4), p.118.

CHAPTER 5

The Eureka Effect

'I like to refer to the septiles as the consciousness-expanding aspects.'

Delphine Jay[1]

FROM THE DATE WHEN TYCHO BRAHE dismounted from his horse to behold a galactic supernova in 1572, to the discovery of DNA structure by James Watson in 1953, spans four centuries of scientific endeavour. We were able to find a total of twenty-three dated Eureka moments over this period - the most comprehensive collection ever assembled, more than twice as many as Koestler described. We have not muddled them up with invention and discovery moments as others normally do, but have limited them to genuine flashes of intellectual insight. Every one of them is dated, which is the best criterion for separating real from mythical E-moments: however vividly a mythical E-moment may be described, it will never have a date.

Table 5.1 shows our list of 23 E-moments, with their time and date, and each with their pair of quintile and septile aspect scores on the right (as will be discussed later). The events for which we have some time-of-day information as well as date are given in italics.

The scientist concerned had to feature in Asimov's *Biographical Encyclopaedia of Science & Technology*, (referred to hereafter as *BEST*), thereby excluding marginal and possibly trivial E-moment cases, and helping to preserve the definition of the E-moment as Koestler described it in his *Act of Creation*. The list has had just a

Table 5.1: The Eureka moments (n=23)

Date	Time (GMT)	Name	Discovery	Q+S
1572 Nov 21	18:00h	Brahe	*Supernova*	3,2
1610 Jan 7	17:00h	Galileo	*Jupiter moons*	4,2
1618 May 15	13:00h	Kepler	3rd law	1,4
1807 Oct 6	14:00h	Davy	Potassium	1,4
1831 Aug 29	14:00h	Faraday	Electromagnetism	3,4
1838 Sep 28	14:00h	Darwin	Natural selection	4,5
1846 Sep 23	23:20h	Galle	*Neptune*	3,5
1869 Mar 1	14:00h	Mendeleef	*Periodic Table*	0,5
1878 Sep 8	14:00h	Edison	Electric light	4,3
1895 Mar 23	10:00h	Ramsay	Helium	7,2
1895 Nov 8	23:00h	Röntgen	X-rays	2,7
1896 Mar 1	14:00h	Becquerel	Radioactivity	1,2
1915 Nov 18	13:00h	Einstein	Relativity	4,3
1921 Mar 28	02:00h	Loewi	*Nerve transmission*	2,3
1923 Oct 6	07:00h	Hubble	Galaxies	3,3
1925 Jun 8	13:00h	Heisenberg	Quantum mechanics	4,2
1928 Sep 3	09:50h	Fleming	*Penicillin*	4,7
1930 Feb 18	23:00h	Tombaugh	*Pluto*	3,1
1933 Sep 12	13:00h	Szilard	Chain Reaction	3,4
1934 Oct 22	12:00h	Fermi	Slow neutrons	0,1
1938 Dec 24	09:00h	Meitner & Frisch	Atomic fission	1,4
1951 Apr 26	12:30h	Townes	*Laser*	3,7
1953 Feb 28	10:00h	Watson	*DNA*	1,4

		Quintile	Septile
Total quintiles and septiles:		61.0	84.0
Chance-expected totals:		50.4	55.4
Percent Excess:		21%	52%

Timed moments in italics

couple of items added since we originally published it in 1988 (those of Edison and Hubble). From comments and discussion it seems likely that only one of the moments here featured was erroneous: the night when Clyde Tombaugh discovered Pluto. He was sifting mechanically through thousands of photographic plates when his big 'aha!' moment arrived (see chapter 6), but this was hardly a moment of insight. The event has a rather low score of (3,1) and one would not wish to appear to be enhancing the significance of the group through excluding this event.

There is also slight doubt concerning one of the earliest e-moments, when Galileo first observed the moons of Jupiter. The night of January 7th, 1610 was certainly a memorable one for the history of astronomy, when a group of stars around Jupiter, never before suspected, were discovered. A telescope was first pointed at Jupiter, and on that very night Galileo wrote off a letter to a colleague about this great discovery. As Galileo's *Starry Messenger* makes clear with its night-by-night account of events, it was not until some nights later that he realised that the new stars were satellites orbiting Jupiter. The experts do not seem clear as to which was the night on which this realisation came to him.[2]

Thus, one or possibly two members of this group of 23 may be a little doubtful, but otherwise persons inspecting the list have felt that it was a genuine collection of E-moments. Astrology author and philosopher Dr Theodore Landscheit who has studied some eureka experiences, wrote to us that 'All events in the list are, in my opinion, genuine scientific eureka moments, and as far as I could see, there are no contradictory statements as to the dates.' One looks forward to further discussion of this matter, and whether any other events could or should have been included. A brief summary of these moments follows, to help us get a feel for these moments, then the next chapter will go into more detail.

The Dated Eureka Moments

1572: Tycho Brahe discovered a new star, 'Tycho's star', a galactic supernova which flared brightly for a year. Tycho Brahe published an account of it which established his name. The previous nova in the 11th century seems to have passed unrecorded in Europe because the heavens were believed not to change. Brahe's discovery must have been 'between 6.00 p.m. and 8.00 p.m.' according to astronomy author Norman Davison, on the grounds that Brahe was on his way to have supper when he had seen the star 'high overhead'.[3] Astonished, he dismounted from his horse and asked his companions whether they, too, could see the new star. They could.

1610: Galileo Galilei saw the moons of Jupiter using his telescope — 'a memorable night in the history of astronomy'[4] — and communicated the observation in an anagram. He studied them an hour and a half after sunset according to J. Meeus, a Belgian astronomer, who computed when the moons would have been in the position drawn by

Galileo. In his notebook for January 7th, 1610, he wrote that he first saw Jupiter in 'the first hour of the night', and that he then 'became aware of 3 little stars ... lying near it.'[5]

1618: Johann Kepler discovered the 3rd law of planetary motion. 'I feel carried away and possessed by an unutterable rapture', he wrote.

1807: Humphrey Davy prepared potassium using electrolysis of molten potash. This was the first of some ten new elements he prepared, one-third of those then known. His biographer, Knight, says that Davy 'danced around the laboratory when he finally succeeded in separating the globules,' of potassium.[6]

1831: On August 29th, Michael Faraday found out how to make an electric current by moving a magnet, described by Asimov as 'probably the greatest single electrical discovery in history.' After ten days of hard work his paper on *The Induction of Electric Currents* was ready.[7]

1846: At the Berlin Observatory, Johann Galle discovered Neptune shortly after midnight on September 23, tracking the object until it set at 2.30 a.m.[8]

1869: Dmitri Mendeleef envisioned the Periodic Table while composing a chemistry textbook. One day he awoke from his afternoon nap with the basic idea, and by the evening he created the first Periodic Table of elements.[9]

1877: Thomas Edison conceived the phonograph principle while using a Morse code tapper. From a device transmitting electrical messages, he conceived one to transmit sound waves by oscillation: 'A sudden reversal of logic and the phonograph was born.'[10]

1895: William Ramsay ascertained that gas from a rock sample was helium, the element previously seen only in the Sun's corona.

1895: Wilhelm Röntgen discovered X-rays just before midnight, while experimenting with cathode-ray tube emissions.

1896: Henri Becquerel discovered radioactivity from uranium ore, after leaving it with a photographic film in a drawer for several days. He noticed a darkening of film on removing it.

1915: Albert Einstein realised that Mercury's orbit could be accounted for by his relativity theory. 'This discovery was, I believe, by far the strongest emotional experience in Einstein's scientific life, perhaps in all his life ...' wrote his biographer. 'For a few days

Chapter 5. The Eureka Effect

I was beside myself with joyous excitement' Einstein said, of that moment.[11]

1921: Otto Loewi discovers the principle of nerve transmission: suddenly awaking at 3 a.m. with the idea of chemical-electrical transmission of nerve impulses, he 'rushed into his laboratory' to check it.[12]

1923: Edwin Hubble was examining an image of the Andromeda nebula at the Mount Palomar telescope one morning, when it dawned upon him that it had to be a separate galaxy outside the Milky Way: this was 'the night the universe changed forever', as his biographer described it.[13]

1925: 'I was shocked to the core', recalled Werner Heisenberg, of the time when the principle of quantum mechanics dawned upon him: 'I had the feeling that I was seeing through the surfaces of atomic phenomena to their deep underlying basis, which had a remarkable inner beauty. So excited was I that I could not think of sleeping, but remained awake all night and watched the sunrise.'

1928: Alexander Fleming discovered penicillin, after a spore entered a dish he was using and stopped a mould from growing. In Koestler's view, 'Fleming had been waiting for that stroke of luck for 15 years; and when it came he was ready for it .'[14]

1930: Clyde Tombaugh found Pluto, supposedly predicted from perturbations of Neptune's orbit. He found it by time-lapse photography, at the Lowell observatory.[15]

1933: Leo Szilard realised how a nuclear chain reaction could take place, while crossing a road in Holborn in 1933. He had just read a newspaper account of a lecture by Lord Rutherford, declaring that any notion of obtaining energy from the atom was the 'merest moonshine.'

1934: It dawned upon Enrico Fermi that, for neutrons to enter the atomic nucleus, they first had to be slowed down by passing through a suitable medium. This he recalled as 'probably the most important discovery that I have made.'[16]

1938: Lise Meitner comprehended fission of the uranium nucleus while walking in the snow in Sweden, after she received a letter from Otto Hahn describing his experiments.

1951: Charles Townes envisioned the laser principle involving coherent oscillation of molecules, while sitting on a park bench in Washington early in the morning. This is our best-timed E-moment.

1953: James Watson realised that the four base-pairs of the DNA molecule could fit together by hydrogen bonds to give a helix structure.[17]

The Natal Group

As well as these dated eureka-moments, there are other equally celebrated experiences that are undateable, for example the flash of insight whereby the principle of the alternating-current electric motor dawned upon Nicola Tesla in Budapest Park. From this quite large group of dated plus undated E-moments, we derive the *natal group* — famous scientists who have recorded E-moments, and whose birth data is reliably known. They have to be in the Asimov reference[18] — as includes over a thousand thumbnail-sketches of famous lives — and have English-language biographies. The latter condition is essential for evaluating the E-moment. This generates the first column in table 5.2, the list of sixteen *eureka-scientists*.

Nowadays, astrological data collections are sold as CD-Roms and stored on websites,* whereas when we were compiling these groups in the 1980's there were various books of data, such as those of the Gauquelins.[19] By general consent, the best CD-Rom on eminent persons birth data is that of Lois Rodden, and this will produce a sub-group of eminent 'scientists and inventors.' Selecting out the reliable (i.e. with a decent source) data from this group, then finding which of these feature in Asimov's *BEST*, will more or less give one the members of table 5.1. The reader may prefer to use more modern books of famous scientists than the Asimov reference, now out of print; but, these may not give the dates of birth as did Asimov, as

---

*The Lois Rodden data collection remains the authoritative one. It classifies data sources into grades of diminishing reliability A, B, C and D, from A implying sight of birth certificate to DD as being 'dirty data' with conflicting data sources, while 'C' means there is no definitely-known source. Her classification has become widely accepted. Project Eureka used material of Rodden grades A and B, and of those given by the Rodden CD-Rom under the label 'scientist' of these grades, about seventy feature in Asimov's *BEST*. Fortunately, the Rodden database is being maintained at http://www.astro.com by Astrodienst, as an 'AstroDatabank Wiki Project'.

was handy, plus he had a penchant for citing any notable eureka-moments enshrined in the biographies, as was a great help.†

That's the easy bit. Then, one has to seek out the biographies, by rummaging through libraries,‡ and more recently by computer-searching the British Library in London. Thereby, the total set of {eminent scientists of reliable birth data, with English-language biographies} comes into focus, as comprises table 5.2. We'll allude to the columns of this table as the E- and non-E natal groups. They have at present 16 and 20 members respectively. These numbers may seem small, reflecting no doubt the traditional antipathy between science and astrology, as well as the fact that Britain is the one country in Europe as does not put time of birth upon the birth certificate.

So, we divide the natal group into two by whether or not the individuals remembered one moment of inspiration as having been crucial for their future life. It is hardly necessary to add, that this concept is unfashionable amongst science historians. Making that decision sets up a complimentary non-E group, where *no eureka experiences were described in their biographies*. This doesn't necessarily mean that they never had one. The non-E group may tend to be more Saturnine and less inspirational or 'Uranian' than the E-group, and, yes, its members may be a little less eminent.

In the earlier publications and presentations on the 'Eureka Effect' we ignored the non-E group in our analyses.§ It then was rather small, only about half the size it is now. It has since grown in size,

---

†The use of Asimov's *BEST* seems to have lost us only one notable figure of known birth time, the computer genius Alan Turing. Two biographies exist of him, neither hinting at any E-moment. A military secrecy shrouding his later life has impaired his reputation until recently.

‡London libraries at the Wellcome Institute and Royal Society are public-access with excellent scientific-biography sections.

§This table was published in 1988 by myself and Mike O'Neill, being much the same as we gave in 1986 at the first presentation of 'Project Eureka' at the Astrological Research Conference, London, and it was republished in 1996 without alteration. Appendix C gives the original list, as presented in 1986. 'The Eureka Effect, a initial Report' by N.K. & M.O. was self-published in 1988. We also published a similar 'Eureka Effect' text in *The Astrological Journal*, 1988, 2, pp.90–97, and 3, p.136; then later, expanded somewhat as a booklet by the Urania Trust, 1996. Since then, the birth data of two more E-scientists has become available, namely Wilhelm Röntgen and Werner Heisenberg, so they have been added to the list. Appendix A gives the data-sets.

partly because of the computer retrieval processes described, which has made it easier to find the reliable natal data plus the biographies, and partly because of the new biographies published since then. Let's not regard these lists as cast in stone. What matters, I suggest, is that one is only adding and not removing individuals from these lists, as the latter could be suspicious. At its present size, the non-E group appears to have just as significant a story to tell as the E-group, but in the opposite direction, which is a surprise and quite unexpected.

Table 5.2: Famous scientists of known birth time in chronological sequence by date of birth with quintile and septile scores.

Eureka	Quintile	Septile	Non-Eureka	Quintile	Septile
Brahe	3	7	Copernicus	1	1
Galileo	2	4	Vesalius	1	1
Kepler	5	3	Hooke	2	1
Davy	2	6	Flamsteed	2	1
Pasteur	4	0	Halley	1	1
Röntgen	1	3	Lavoisier	4	2
Edison	0	6	Bode	1	1
Becquerel	3	2	Brewster	4	4
Tesla	2	3	Leverrier	1	1
Einstein	4	3	Huxley	3	0
Heisenberg	0	2	Crookes	3	2
Fleming	5	2	Curie	1	1
De Broglie	4	3	Hahn	2	2
Fermi	2	3	Joliot-Curie	1	0
Townes	2	3	Pauling	1	0
Watson	4	6	Segré	4	1
			Bethe	1	3
			Seaborg	2	3
			Ehrlich	3	0
			Sagan	1	3
Total Score:	43.0	56.0	Total Score:	39.0	28.0
Expected:	35.0	38.6	Expected:	43.8	48.2
% Excess:	+23	+45	% Excess:	−11	−42

	Quintile	Septile
Bell	3	2
Marconi	2	1

We required that the natal group data be reliably timed — in contrast with the E-moments. It seemed likely that, especially prior to the advent of the electric light, the time of insight was likely to be within several hours of 2.00 p.m. That would make its uncertainty much less than plus or minus twelve hours which is the possible error for untimed birth data. We, therefore, accepted any E-moments as long as they were dated.

We were able to place all of our group of Asimov scientists into one of two categories, except only for Alexander Bell and Guiglio Marconi. These two both had remarkable 'aha!' moments, but they were what we later came to call invention moments, times of concrete achievement, for the telephone and radio, respectively.

The term 'scientist' is a broad one, and we endeavoured to use it in a way that would promote consensus. The philosophers Francis Bacon and René Descartes, the psychoanalyst Sigmund Freud, and the artist Leonardo da Vinci, are all in Asimov and of known birth time, yet we excluded them. Science historians discuss Freud's work, and books on invention include it, yet many don't regard psychoanalysis as being a science. Also Samuel Morse, the inventor of the electric telegraph, was not included, because he was by profession an artist.

We chose not to include mathematicians, even though that Henri Poincaré's well-known account of his eureka experience while stepping on a bus is discussed in the *Act of Creation* of Koestler. This leaves a separate group on which our hypothesis can be tested. The mathematicians Euler and Gauss have described such moments in their lives, but it seemed preferable to leave the subject to others better qualified to delve into the history of mathematics. Mathematicians have different characters from scientists and so, for the analysis of planetary aspect frequencies, we preferred to keep them apart.

Birth data had to be reliable, which means that they came from birth certificates or from a close relative of the person, or the person themselves. For example, Clyde Tombaugh who discovered Pluto wrote to us that he was born 'probably in the early morning hours' and this could not be used, though the moment when he discovered Pluto is known and features in table 5.1.

Summarising, the Eureka natal group was selected by four criteria, namely that the persons included had to: be scientists, referred to in Asimov's *Biographical Encyclopaedia of Science & Technology*, with times of birth known reliably to within an hour; and having sufficiently detailed biographies in English to permit evaluation of whether they experienced an E-moment.

Two persons compiled the lists, one having the computer program able to generate the score for a given time (M.O.), and the other (N.K.) a science historian by training who made a yes/no decision *prior to* that score being ascertained. Were any cases transferred across from one group to another after their quintile and septile score had been found? The answer, as stated in earlier presentations, is that for one case only, that of Leverrier, was this done. Initially, August Leverrier was scored as having 'discovered' the planet Neptune. This was a mistake, as there was clearly no sudden insight on his part. He predicted a planet's position, but didn't find it. A sceptic could argue that Leverrier's low score (1,1) influenced the decision to transfer him into the non-E group. Clearly, this key classification decision should be made prior to scoring.

Significance of Results

We can now begin to analyse the data, as shown in table 5.2. The Eureka-group of scientists had 34% more quintiles and septiles than one would expect by chance, and they had 85% more of these aspects per chart than did the non-eureka group. They had a staggering *250% more septiles in their natal charts than did the non-E group.* That is to say, they averaged 3.5 septiles per chart as compared with only 1.4 per chart for the non-E group. This data shows how the *same effect* has replicated through both the natal and event groups.

The hypothesis here tested was that there would be an excess of quintile and septile aspects in the natal charts of eureka scientists. For this we used the orbs as proposed by Addey, namely just over two degrees for quintiles, $\frac{12°}{5}$, and just under two degrees for septiles, $\frac{12°}{7}$. That hypothesis has now been confirmed. We then predicted that the same should be found for the E-moments. To test these predictions, we had to generate the chance or expected values for these aspects. That isn't easy, and involves having the computer select thousands of randomly-selected times and score them. The expected frequencies, that is to say the number as would be expected per chart, of the quintile and septile aspects at these orbs came out at 2.2 and 2.4 over the historical period concerned. Those figures were derived from computer sampling over the four centuries spanned by the list (see appendix B).

These data-sets show a septile effect uniformly stronger or with a larger amplitude than the quintile effect. Let's compare them from the two tables:

Table 5.3: Excess of aspects

Group	Number	Quintiles		Septiles	
E-moments	23	61/50.4	+21%	84/55.4	+52%
E-scientists	16	43/35.0	+23%	56/38.6	+45%
Non-E scientists	20	39/43.8	−11%	28/48.2	−42%

The septile effect is working most strongly in the event-group. It is clear from this that septiles have an illuminative quality, which became dominant at these historic moments. Then, we are startled to notice that the non-eureka group is displaying almost as large an effect as the natal E-group, but in the other direction. The aspects are there in deficit. No-one predicted this in advance. This powerful contrast between the two groups, present in the quintiles but far more strongly in the septiles, must point to a real psychological difference between these two types.

The appropriate test for checking out the significance of these results is the 'chi-squared' and it's about the simplest test known to statistics. For the list of eureka-moments in table 5.1, the excess score of these aspects above chance has a significance level of one part in two thousand (a chi-square of 15¶). At that level of significance chance can be ruled out, i.e. the evidence is conclusive — if, indeed, the list has been properly constructed. Really, these tests only tell us what can be seen by inspection, as to how significant an effect is. The smaller natal E-group has a lower level of significance, of around 1 in 300.

¶In earlier presentations, M.O. expressed concern that the units involved were not quite independent (as the chi-square test requires), because the presence of some of these aspects in a chart could affect the likelihood of further such aspects also being present. He devised an empirical method of assessing significance, as involved randomly generating huge numbers of charts and counting their aspects. Its net result, however, was merely to confirm the significance values earlier obtained using the chi-squared test, see appendix B.

The Non-Eureka Group

Many scientific biographies do not focus on one moment of realisation. Instead they describe the gradual unfolding of a theory or insight. Let's here quote from a biography of Marie Curie written by her daughter Eve:

> The layman forms a theatrical — and wholly false — idea of the research worker and of his discoveries. The 'moment of discovery' does not always exist: the scientist's work is too tenuous, too divided, for the certainty of success to crackle out suddenly in the midst of his laborious toil like a stroke of lightning, dazzling him by its fire. Marie, standing in front of her apparatus, perhaps never experienced the sudden moment of triumph.[20]

Likewise, in the biographies of Copernicus, Halley or T. H. Huxley, we find no memorable date when their perception of a problem changed: if they did have such a moment, it did not get into their biographies. Copernicus left a record of how he 'gradually began to meditate' upon the motion of the Earth.[21]

Thus the 'non-Eureka' group is defined negatively, as that group of eminent scientists whose biographies do not describe such moments. The non-eureka group was developed as a complement to the eureka group, to see to what extent a converse situation applied to them in respect of scores allocated.

As an example, let's quote the first preparation of oxygen gas by the English clergyman, Joseph Priestley, on August 1st, 1774. On that day Priestley tried out an ingenious new method, heating mercury oxide in a retort using a lens to focus the sunlight, collecting the gas produced over mercury. Concerning what he had discovered he wrote, 'When the decisive facts did at length obtrude themselves upon my notice, it was very slowly and with great hesitation, that I yielded to the evidence of my senses.'[22] Later on Priestley came to regard this as a decisive experiment, becoming famous from the discovery due to Lavoisier explaining its significance. Left to himself he would just have called his discovery 'dephlogisticated air' and viewed it as related to the nitrous oxide which he had earlier prepared. According to Thomas Kuhn, Priestley in this experiment believed he had prepared just one in a series of 'airs' that came from heating different solids, i.e. there was nothing resembling an 'Aha!' moment when he twigged that oxygen existed. Priestley first published an experiment to make oxygen: it was first made in 1771 by the Swede Carl Wilhelm Scheele who called it 'fire air.'[23]

The discovery of the planet Uranus is a classic example of a non-eureka process, extending over about six months. Several astronomers had recorded Uranus as a star, and then on March 13th, 1781, William Herschel discovered that this star previously noticed was of a disc-shape. His telescope was more powerful than anyone else's at the time, and he was conducting a systematic sky-search. A few nights later he noted that it was in motion against the stars, and so concluded that it was a comet. He presented his *Account of a Comet* to the Royal Society a month after that. Only about six months later did people gradually apprehend, by following its path in the sky, that it must be a new planet.[24]

Two well-known American 'Asimov' scientists, Hans Bethe and Carl Sagan, have biographies which contain no hint of E-moments,[25] and have reliable birth data. Hans Bethe discovered what is (supposedly) the carbon cycle of nuclear reactions inside the Sun. A colourful account of the eureka experience for this breakthrough was given by the popular science writer George Gamow, according to which the theory came to Bethe on a train journey just as the Sun was setting and dinner was being served.[26] A letter was sent to Professor Bethe about this, and his reply dismissed this account as a mere prank by Gamow, declaring that no such distinct moment had existed for him! His biography reiterated his dismissal of the Gamow story, explaining that only slowly over months, did he work out the equations.[27] Bethe's Q+S score is (1,2).

The biography of the French physicist Frederic Joliot-Curie tells how he collaborated with his wife in the 1930's, that crucial decade following Pluto's discovery, in generating artificial radioactivity. Conveniently, it gives his time of birth. There is no sign of any eureka moment, though it does give the momentous time when they first made an artificial element;[28] rather alchemically, an exact New Moon was then conjunct Mercury and opposing Pluto. His natal Q+S score is (0,1).

The 'Mandelbrot set' whose wondrous forms unfold on today's computer screens, having their roots in the realm of imaginary numbers, was discovered by Benoit Mandelbrot. The images of his 'set' started to emerge from a computer at Harvard University in 1980, when Mandelbrot recalled how 'they lit my life with intellectual and aesthetic revelations.' A letter of his may be quoted (with permission) as to how he first perceived a 'Mandelbrot set' early in 1979 but failed to appreciate its meaning:

> Your query has made me think, and has helped me pinpoint more sharply a permanent feature of my research style. Here is a brief response. It is obvious that I know of scientists who claim to have experienced 'eureka moments'. But I have seldom experienced anything that fits that description ... my first sighting of that set was not in my opinion a clear-cut discovery, but merely one part of a wide complex of observation that eventually spurred me to the work described ...

His thought processes, he explained, had followed a more 'fractal' path ... Mandelbrot is thus a non-eureka type. He is not included in the list for three reasons: his discovery is mathematical, he is too recent to feature in Asimov, and his birth time is unknown.

Predictions from the Theory

A scientific theory needs to make testable predictions. Here are seven ways in which the Eureka Effect here described can be tested.

- Mathematicians such as Euler, Gauss, Hamilton, Boole and Poincaré have described E-moments. André-Marie Ampère (1755–1836), after whom the unit of electric current is named, recorded in his diary the circumstances of his first mathematical discovery, on April 27, 1802:

  It was seven years ago I proposed to myself a problem which I have not been able to solve directly, but for which I had found by chance a solution, and knew it was correct, without being able to prove it. For some days I had carried the idea about with me continually. At last, I do not know how, I found it, together with a large number of curious and new considerations concerning the theory of probability ...²⁹

This seems to have been a purely mathematical insight. Its Q+S score is (3,3). Better known is the E-moment of Sir William Rowan Hamilton, which struck while he was strolling across Brougham Bridge near Dublin. It is well-timed and dated, and described in chapter 12: DNA and The Decile. The group of mathematical E-moments and natal data form a distinct set.[30] [31] [32]

- Can the date be found, of Coleridge's opium-aided inspiration for his masterpiece *Kubla Kahn*, or Mary Shelley's nightmare about Baron Frankenstein and his creation (after an evening spent telling ghost stories around the fire with her husband and Lord Byron), or the day in Paris when the first haunting strains of Tchaikovsky's 'Pathétique' symphony came to him? Frankly, I doubt it. Of the composing of Rilke's Duino Elegies it was said:

   They are named for the castle of Duino, which stands on a rocky headland of the Adriatic above Trieste where, during a lonely winter sojourn in 1912, some of them were first conceived. Ten years of distressing silence followed this promising beginning, and then, in another lonely castle in Switzerland, the spell was broken and the ten elegies took shape. 'All in a few days, it was an indescribable storm, a mental and spiritual hurricane (as in those days at Duino), every fibre, every tissue in me cracked — eating was never to be thought of, God knows what nourished me.'

Could an expert on literature seek out such moments of notable inspiration? He or she would have to lack access to their quintile and septile scores, while doing this.

- John Addey discussed septiles in the context of musicians who were especially inspirational — and others have made similar claims about poets.[33] Addey claimed that their 'seventh-harmonic' charts were 'strong'. If an expert on the subject were to assemble such a group, with their historic moments of inspiration, one could then find the charts and score their quintiles and septiles. Thereby, one would be applying Koestler's view that experiences of inspiration in art and science were comparable.

- There are nearly three dozen scientists in the Asimov reference for whom a reliable birth time is available, but for whom no detailed biographies exist in English (appendix A). This group is a source whereby our theory could be tested.

- A survey published in a chemistry journal found that E-moments were not uncommon in the lives of ordinary working chemists. Recollections such as the following were found:

  Freeing my mind of all thoughts of the problem I walked briskly down the street, when suddenly at a definite spot which I could locate today, as if from the clear sky above me an idea popped into my head as if a voice had shouted it. ... The idea came with such a shock that I remember the exact position quite clearly. ... I decided to abandon the work and all thoughts relating to it, and then on the following day when occupied in work of an entirely different type, an idea came into my mind as suddenly as a flash of lightning and it was the solution ... the utter simplicity made me wonder why I hadn't thought of it before.[34]

If the dates of such moments could be located then they would offer an opportunity for testing the hypothesis here advanced on non-eminent persons. One hopes that scientists will become more careful in remembering these dates.

- Conferences or musical events requiring an inspirational tenor could be scheduled for days when strong septile aspects were present. The Harmogram (see chapter 12: DNA and The Decile) gives an easy way to locate such dates. In retrospect, one could discuss to what extent such had been achieved. Musicians would tend to be more open-minded to such a suggestion than groups of scientists.

- Some decades from now, there should be enough fresh E-moments for a further test to be feasible.

References

1. Delphine Jay, *Practical Harmonics* AFA California, 1983 p.7. Her work is influenced by John Addey's 'Harmonics in Astrology', 1976.

2. Stillman Drake, *Galileo at Work*, 1980, p.152.

3. Letter from Norman Davidson, author of *Astronomy and the Imagination*, 1985.

4. Drake, ref. (2), p.143.

5. Meeus J., Galileo's first Records of Jupiter's Satellites, *Sky and Telescope*, September 1962, p.137.

6. Knight, *Humphrey Davy*, 1992, p.65.

7. Pearce Williams, *Michael Faraday*, 1965, p.165.

8. Patrick Moore, *The Planet Neptune, An Historical Survey before Voyager*, New York 1988, p.23.

9. J. Elmsley, 'Mendeleev's Dream Table', *New Scientist*, March 7, 1985.

10. Koestler *AOC*, p.197; Hughes, 1979, 1101.

11. R. Pais, *Subtle is the Lord: the Science and Life of Albert Einstein*, Oxford 1982, p.253.

12. Koestler, *AOC* p.205.

13. Christianson, H., 'The Night the Universe Changed Forever.' *The Griffith Observer*, June 1997, pp.4–10.

14. Koestler, *AOC* p.194.

15. Moore, P. and Tombaugh, C. *Out of the Darkness — the Discovery of Pluto*. New York 1984. Tombaugh wrote to us that he detected Pluto on photographic plates at 04:00hrs ±3 mins, Mountain Standard Time.

16. E. Segré, *Enrico Fermi, Physicist* Chicago 1970 p.81.

17. Olby R., *The Hunt for the Helix*, 1974, p.412. A letter from J.D. Watson confirmed the time.

18. Isaac Asimov, *Asimov's Biographical Encyclopaedia of Science & Technology* 1966, New York.

19. Gauquelin M. & F., *The Gauquelin Book of American Charts*, ACS, CA 1982.

20. E. Curie, *Marie Curie*, 1931, p.165.

21. Koestler, *The Sleepwalkers*, p.207.

22. J.R. Partington, *A History of Chemistry*, III, 1964, p.265.

79

23. Partington, op. cit., Vol. III, p.219.

24. Thomas Kuhn, op. cit., p.115.

25. Keay Davidson, *Carl Sagan a Life* 1999 (Our inclusion of Sagan into the non-E group in 1987, prior to this biography's appearance, was not strictly legitimate.)

26. G. Gamow, *The Birth and Death of the Sun*, 1940, p.113.

27. Koestler, ref. (21), p.46.

28. M. Goldsmith, *F. Joliot-Curie, a Biography* 1976.

29. Koestler, *AOC* p.117.

30. Jacques Hadamard, *The Psychology of Invention in the Mathematical Field*, Princeton, N.J., 1945.

31. Marie Louise von Franz, *Number and Time*, 1974, Chapter Two 'Images and Mathematical Structures ... '

32. There are some mathematical E-dates in *The Penguin Dictionary of Curious and Interesting Numbers*, D. Wells, 1986.

33. Charles Graham, 'The Seventh Harmonic and Creative Artists', *Astrology Now*, 13 June 1976, pp.58–89.

34. W. Platt and R.A. Baker, 1931, 'The Relation of Scientific Hunch to Research', *Journal of Chemical Education*, 8, 1969.

CHAPTER 6

The Moment of Illumination

'Moved with delight, he leapt out of the pool'

Vitruvius on Archimedes

ONE DAY THE HUNGARIAN ÉMIGRÉ PHYSICIST Leo Szilard was crossing a road, when the notion of a nuclear chain reaction dawned upon him:

> In London, where Southampton Row passes Russell Square, across from the British Museum in Bloomsbury, Leo Szilard waited irritably one gray Depression morning for the stop-light to change. A trace of rain had fallen during the night; Tuesday, September 12, 1933, dawned cool, humid and dull. Drizzling rain would begin in early afternoon. When Szilard told the story later, he never mentioned his destination that morning. He may have had none; he often walked to think. In any case another destination intervened. The stop-light changed to green. Szilard stepped off the curb. As he crossed the street time cracked open before him and he saw a way into the future, death into the world and all our woe, the shape of things to come.

This can safely be described as the most dramatic account of a man crossing a road in all literature. As the opening paragraph of Rhodes' weighty opus, *The Making of the Atomic Bomb*, it portrays a classic eureka experience.[1] This is how Szilard himself described the event:

> It suddenly occurred to me that if we could find an element which is split by neutrons and which would emit two neutrons when it absorbs one neutron, such an element, if assembled in sufficiently large quantities, could sustain a nuclear chain reaction ... the idea never left me.

Just one Jupiter cycle after this experience came the events of Hiroshima and Nagasaki: Jupiter's position at these cataclysmic events was conjunct its position at Szilard's E-moment. The potential of those events somehow lay within that E-moment of Szilard, as Rhodes has indicated.*

Our E-moments do not all have the sudden lightning-bolt quality of Szilard's account. We here describe a dozen such, selected from the list of twenty-one already given. Following Koestler's approach, we stress the inner or subjective side of the experience, and show how this enabled the decision about date and time to be reached. Location of the date can involve difficult detective work. In the above case it was reconstructed from the date of a speech by Lord Rutherford at a scientific conference, when he categorically denied that energy could come out of the atom. That was the day before Szilard's experience, since it was the reading of a newspaper account of that speech that stimulated him to think otherwise.

Kepler's Third Law

1618, 16 MAY   1Q, 4S

The poet Coleridge much admired Kepler's discovery of his third law of planetary motion (which states that the cubes of the orbit periods are proportional to the squares of their mean distances from the Sun) and he declared:

> There is not a more glorious achievement of scientific genius upon record, than Kepler's guesses, prophecies and ultimate apprehension of the law of the mean distances of the planets as connected with the periods of their revolutions round the Sun.[2]

---

*For Szilard's role in operating the first nuclear chain reaction in Chicago in 1942, together with Enrico Fermi, see chapter 9: When Inventions Worked.

Chapter 6. The Moment of Illumination

In 1617, Kepler came across Ptolemy's *Harmonia*, which brought him 'an especial increase of his passionate desire for knowledge and encouragement of his purpose.' His biographer Caspar described how 'Kepler in those weeks and months must be pictured in a condition of greatly increased excitement, such as seizes an artist who has an idea which strives to be shaped.' (p.267, *Harmonia*). It was then that Kepler wrote, 'I feel carried away and possessed by the unutterable rapture of the divine spectacle of the heavenly harmony.'

An account of this experience was given in Kepler's *Harmonices Mundi* in a section entitled, 'The Main propositions of astronomy necessary for the contemplation of the heavenly harmonies':

> After I had discovered the true intervals of the orbits by ceaseless labour over a very long time and with the help of Brahe's observations, finally the true proportion of the periodic times to the proportion of the orbits showed itself to me. On the 8th March of this year 1618, if exact information about its time is desired, it appeared in my head. But I was unlucky when I inserted it into my calculation, and rejected it as false. Finally, on May 15th, it came again & with a new onset conquered the darkness of my mind, whereat there followed such an excellent agreement between my 17 years of work at the Tychonic observations & my present deliberation that I at first believed I had dreamed and assumed the sought-for result in the supporting proofs. But it is entirely certain and exact that the proportions between the periodic times of any two planets is precisely one and a half times the proportion of the mean distance.

Inspiration struck against a stormy background, with the Thirty Years War breaking out the following week, and his mother being put on trial for witchcraft that May. His triumphant account written shortly after the discovery on May 15th, 1618 recalled how:

> Now, because 18 months ago the first dawn, three months ago the broad daylight, but a very few days ago the full Sun of a most highly remarkable spectacle has risen, nothing holds me back. Indeed, I give myself up to sacred frenzy. I sneeringly defy mortals by the following public

> avowal: I have plundered the golden vessels of the Egyptians in order to furnish a sacred tabernacle for my God out of them far from the borders of Egypt. If you pardon me, I am happy. If you are angry with me, I bear it. Well then, I cast the die and write a book for the present time, or for posterity. It is all the same to me. It may wait a hundred years for its readers, as God also has waited six thousand years for an onlooker.[3]

The third-harmonic was strongest at this climactic E-moment, with a grand trine (a triangle of trines) present in the sky. After all, this was his third law whose formulation includes use of a *cube* power. Kepler was renowned for the way in which his theories involved the number three, often employing the notion of the Trinity in some variant. In his first book, *Mysterium Cosmographicum* he declared that,

> My ceaseless search concerned primarily three problems, namely the number, size, and motion of the planets — why they are just as they are and not otherwise arranged. I was encouraged in my daring inquiry by that beautiful analogy between the stationary objects, namely, the sun, the fixed stars, and the space between them, with God the Father, the Son and the Holy Ghost. I shall pursue this analogy in my future cosmographical work.

The harmogram is a graph giving a flowing image of the strength of a celestial aspect. The one shown in figure 6.1 is of trine-power, i.e. the third-harmonic, and it covers twelve days either side of the event. There were four septiles over this E-moment; if one plots both trines and septiles together, so the two are added, a strong peak at Kepler's E-moment appears, the biggest over a fifty-day period, with that moment at the centre. This combination of 3rd and 7th harmonics peaks in the morning of 15th May 1618, indicating why his brainwave arrived at just that moment.

Davy Discovers Potassium
1807, 06 OCT   1Q, 4S

When the young Humphrey Davy discovered potassium, he became so excited that he almost died. Many had surmised as to what

Chapter 6. The Moment of Illumination

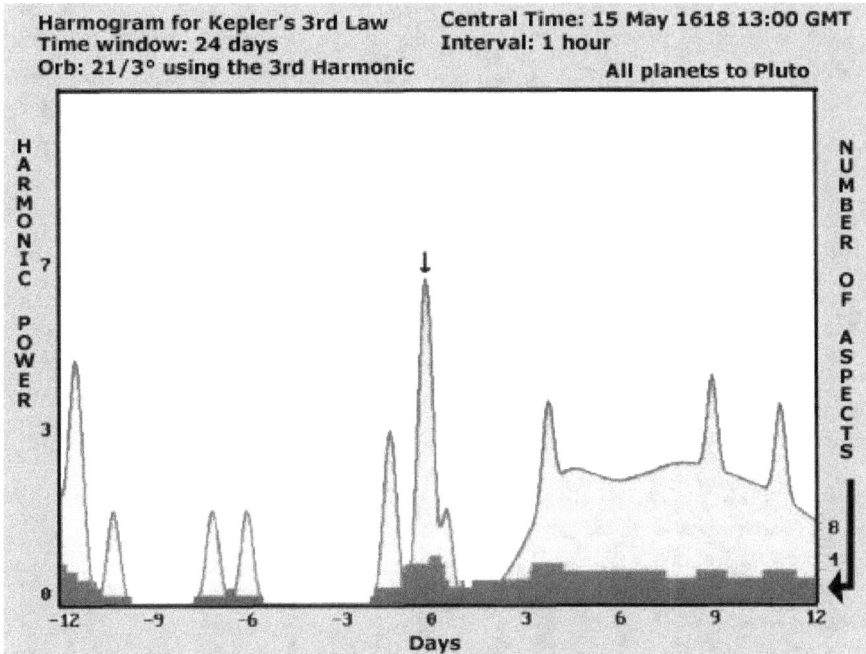

Figure 6.1: Kepler's Third Law Eureka moment (third harmonic)

'potash' might be composed of. We now refer to it as potassium nitrate, but it was then referred to as vegetable alkali, since it came from plant ash. Was it perhaps made of lime and hydrogen, or perhaps phosphorus and nitrogen, as Davy himself suspected? He wondered whether some new element might appear from electrolysing it, and assembled a huge electric battery for the purpose.

Davy first passed a current through a solution of potash, but that didn't work and he then tried fused potash. His assistant recalled how:

> ...when he saw the minute globules of potassium burst through the crust of potash, and take fire as they entered the atmosphere, he could not contain his joy — he actually danced about the room in ecstatic delight; and some little time was required for him to compose himself sufficiently to continue the experiment.

This new metal on touching water 'instantly burnt' and swam on the surface, becoming potash again, in solution.

A few days later, he likewise discovered sodium by electrolysing soda. 'Never, perhaps, was a chemical investigation more intensely interesting than the one under discussion; and never, perhaps, in so short a time were so many new and surprising facts developed' wrote Davy's brother in his biography. Davy realised that potash and soda were metallic oxides. Chemistry was never the same again. The date of 6th October comes from Davy's lecture announcing the discovery.

Davy was a well-known figure from his discovery of 'laughing gas' (nitrous oxide) which had become fashionable at parties. His fame and popularity were much enhanced by these sensational discoveries. 'This was a golden period in his life' to quote his biographer, but added, of the lecture where he reported his great discoveries:

> ...when he composed it he was in a feverish state — the prelude to a severe attack of illness, which was near proving fatal, — and his great apprehension was, that he should die before he had published his discoveries.

Davy's physicians surmised that his illness was brought on by the over-excitement from his experimental labours and discoveries. Bedridden for months, he struggled between life and death. All ranks of society inquired as to his health. He composed a long poem during his fever about how the divine power worked through the frame of nature. Potassium was the first of ten new elements he discovered.

Darwin and Natural Selection
1838, 28 SEP   4Q, 5S

A theory of evolution by 'natural selection' was formulated almost independently by two different people. On June 3rd, 1858, Charles Darwin received a letter from Alfred Russell Wallace, posted three months earlier from a South Sea island. 'Without question he was stunned as he had never been before', to quote Wallace's biographer.[4] It was two weeks before Darwin showed the letter to anyone or mentioned that he had received it. He then confessed that 'all his originality was smashed,' and that the very chapter headings of his embryonic work appeared in the letter from Wallace (See Ch. 6).

Darwin's Eureka-moment had been described by one author in terms of Arthur Koestler's 'bisociation' concept:

Chapter 6. The Moment of Illumination

Darwin was much struck by the infinite number of species and varieties of living forms and how they seemed to vary from one another by degree as though descended from a common source. Yet he was unable to account for the principle which underlay the assortment of creatures. He recounts that it was the reading of Malthus and the gloomy image of the constant struggle for existence which made all the parts of speciation fall into place. In other words, he bisociated the account of human population dynamics to the problem of animal and plant variability, thereby laying the foundation for *The Origin of Species*.[5]

Malthus' essay had been published in 1797, more than forty years earlier. It was a pessimistic work, as to how populations would always grow exponentially, while the means of supporting them would only increase more gradually, so that poverty and the struggle for survival would always be with us.[6]

'I can remember the very spot in the road, whilst in my carriage, when to my joy the solution occurred to me.' recalled Darwin.[7] Darwin's autobiography gave no hint of an E-moment, but emphasised how slowly he had arrived at his ideas.

However, the US Eureka-ologist Howard Gruber made a detailed study of Darwin's notebooks, and concluded that:

> [Darwin] did indeed on 28th September 1838 have one great insight in which he first saw clearly the theory of evolution through natural selection ... his notebooks show that he had or almost had the same idea a number of times before, during the fifteen months of deliberate effort leading up to the moment in question. So the historic moment was in a sense a re-cognition of what he already knew or almost knew ... After having the idea, Darwin reverted to other preoccupations. It took him about two more months for the idea of evolution through natural selection to begin to dominate his thinking ...

It is indeed rare to have an academic analyse a eureka experience. Let's quote further from him:

> ... the insight of 28th September 1838 has the earmarks of a complex, internally structured event. As Darwin

recorded it, at least three modalities of thinking were entailed: visual imagery, internal dialogue, and logical reasoning — as well as the transformations of ideas from one to another of these modalities. The case of Darwin is a prime argument against the 'one great insight' notion. His notebooks are studded with excited passages; he had the same or similar insights repeatedly; he had to have a number of quite different novel ideas and co-ordinate them in a coherent theory; and whatever transformative moments occurred did their work slowly, thereby expressing their complex inner structure.[8]

Gruber was arguing against the notion that E-moments occurred suddenly and out of the blue. This E-moment occurred at his brother Erasmus' house in London near Baker Street, while reading the latter's copy of Malthus, Gruber advised this writer.

In the year 1854, after Wallace had read Darwin's *Journal of a Naturalist's Voyage on the Beagle* and Malthus' *Essay on Population*, he published an article about it in a scientific journal. He postulated that 'every species has come into existence coincident both in space and time with a pre-existing, closely allied species', so that they formed 'a branching tree'. Charles Darwin read the paper and wrote to Wallace that he agreed with 'almost every word' in it, adding that he himself had been working for twenty years on the problem and had a 'distinct and tangible idea of its solution'. Thus the two theories were not quite independent: Darwin encouraged Wallace, the year before the latter's theory occurred to him.

Alfred Russell Wallace left what turns out to be a rather bogus account of an E-moment, adjusting his memory of space and time. To quote his biographer, 'He chose not to do this however [describe where he really was], and instead altered the actual story, thus establishing the legend we all know.'[9] Wallace realised that the idea of a lifetime had come to him, and decided to embellish the story somewhat. In the jungles of Ternate, few would contradict him. While nearly dying, lying in a hammock in the jungle with a fever, Wallace claimed to have recalled the work of Malthus which he had read years earlier:

> Vaguely thinking over the enormous and constant destruction this implied, it occurred to me to ask the ques-

Chapter 6. The Moment of Illumination

tion, 'Why do some die and others live?' and the answer was clearly that on the whole the best fitted live ... Then it suddenly flashed upon me, that this self-acting process would improve the race ... the fittest would survive. Then at once I seemed to see the whole effect of this ... I waited anxiously for the termination of my fit so that I might at once make notes for a paper on the subject. The same evening I did this pretty fully, and on the two succeeding evenings wrote it out carefully in order to send it to Darwin by the next post, which would leave in a day or two.[10]

No copy of that original theory plus accompanying letter survive, presumably destroyed in anguish by Darwin.

Wallace was on the virtually unknown island of Gilolo at the time, in February 1858 when the theory came to him, and was reflecting on how the natives of Gilolo differed in appearance from the Malayans. From January 20th until March 1st he remained on this island, as his published narrative *The Malay Archipelago* indicates, and was sick with fever. Ternate was an exotic spice island which Francis Drake had visited, and clearly a more appropriate spot for the E-moment. The Dutch steamer which took Wallace's letter departed from Ternate on March 9th, 1858. This is verified by a couple of other letters that Wallace sent off by the same post, which do survive. He could not have posted his letter on a boat which left two days later as he declared, because he returned to Ternate from Gilolo, where he formulated his theory, on March 1st.

This tends to suggest that no E-moment really occurred to Wallace — after all, if the circumstances have been so adjusted then why credit the story? Darwin and Wallace became lifelong friends, after their joint paper on the subject was read out to the Linnean Society in July, 1858.

Galle Sees Neptune

1846, 23 SEP   3Q, 5S

'The most magical predictive-math event in the history of the oldest science.'

Dennis Rawlins, on Neptune's discovery[11]

Uranus was discovered in 1781. Four decades later, the French astronomer Bouvard announced he had detected a discrepancy in its orbit, and this became recognised as a disturbing anomaly. Synchronously enough, Uranus and the not-yet-discovered planet beyond it, Neptune, were conjunct in that year 1821 when he made that claim. Finally, two decades later, the mathematician/astronomer Urbain Leverrier in Paris computed the position of a new planet. It was an extraordinary labour of precision and perspicacity, and he gave his first report on the subject in November 1845 to the Paris Academy, and produced a memoir in the following June, and was at once published.

In July of 1846, Britain's Astronomer Royal, Sir George Biddell Airy, received a letter from Leverrier in France, which so impressed him with its exact co-ordinates and air of confidence, that he authorised the search for Leverrier's predicted planet, using the University of Cambridge telescope — without informing Leverrier he was doing this. The search was a dismal failure. A later chapter looks at this failed Neptune-quest, in terms of the synastry (i.e. connection between two charts) between the charts of Urbain Leverrier and John Couch Adams, the British mathematician who was also computing its position.

Leverrier could not arouse much interest amongst French astronomers, and eventually decided to approach the Berlin Observatory. He admired the work of the German astronomer Johann Galle at that observatory, and wrote to him. The Berlin Observatory had freshly-prepared star-maps, the best in Europe, and that for Aquarius had recently been completed.

Neptune was discovered at its conjunction with Saturn, and was, amazingly, *within a degree* of where the Frenchman had predicted it. Saturn and Neptune met thrice that year, the second as they both went retrograde, conjoining at 26° 26′ Aquarius, within arc minutes of Leverrier's Mercury at 26° 31′ of Aquarius! That was in the beginning of September. Leverrier sent off his historic letter on September 18th, 1846 to the Berlin Observatory which was to lead directly to the discovery of the new planet. That day saw some major Neptune aspects:

Chapter 6. The Moment of Illumination

Saturn	conjunct	Neptune	0° 34'
Moon	opposite	Neptune	5° 00'
Pluto	sextile	Neptune	0° 34'

Sending off a scientific letter about Neptune was appropriate on such a day. Jupiter was transiting Leverrier's ascendant (2°), apt for the moment which brought him immortality.

On the night of September 23, 1846, Galle and his assistant decided to take a look. Galle called out the star positions, and D'Arrest verified them on the chart. After one hour, D'Arrest exclaimed: 'That star is not on the map!' They had found a new star less than one degree from where Leverrier had predicted. They located it at a quarter past midnight. It was the most brilliant prediction in astronomy's long history.

On that historic night, when Johann Galle turned the Berlin Observatory telescope towards the corner of the sky specified by Leverrier, Jupiter in the sky was crossing over the New Moon in his horoscope to 2° orb, and loosely opposite his Neptune (5°). His chart shows quite a strong synastry with Leverrier, whom he had never met:

Table 6.1: Synastry between Galle and Leverrier

Galle	**Aspect**	**Leverrier**	**Orb**
Moon	conjunct	Ascendant	2°
Sun	square	Sun	2°
Mercury	square	Mercury	1°
Venus	opposite	Venus	1°

The two were born a year apart, so that all their outer planets were conjunct, e.g., Neptune conjunct Neptune 37'.

The new planet turned out to have a distinct sea-green colour. Leverrier decided to name it after the sea-god Neptune, without knowing that. The words of the poet written 26 years earlier had been fulfilled:

> 'Or like some watcher of the skies,
> When a new planet swims into his ken.'

(Keats, *Lamia*, 1820)

That was the sole occasion in history when such an experience took place, and it did so two decades *after* the poem had been composed! The discovery of Uranus had of course been the stimulus for Keats' immortal lines, but there was no such moment in its discovery: only gradually did mathematicians apprehend that the 'comet' which William Herschel had reported must be a new planet beyond Saturn, as they saw its orbit. Thus eureka-type images tend to appear as ornaments to the historical record.

Some have criticised the inclusion of Galle's discovery of Neptune as an E-moment. It would, however, be sad if our definition of E-moments were not to include that poetic image, replete with wonder and amazement. The event had the unexpectedness of E-moments, and saw two quite different kinds of information suddenly coming together, namely a theoretical prediction based on the inverse-square law of gravity, and the most accurate star-maps in the world.

The first sighting of Neptune had been made much earlier, by Galileo, as Jupiter passed it while he was studying the Jovian moons. Galileo was much puzzled as to what it could be. Galileo's original manuscripts are now in the hands of the Vatican (ironically enough, considering his attitude towards them) and they show that some hand has endeavoured to erase his early location of what was in fact Neptune — supposing that the record must have been a mistake on Galileo's part. Thus Neptune's earliest depiction is now a blurred and half-erased ink stain. Very appropriate, astrologers would remark.

Mendeleef's Periodic Table
1869, 01 MAR   0Q, 5S

In 1864, John Newlands, an English chemist, discovered his 'law of octaves' as he called it. The chemical elements, when arranged in sequence by atomic weight, had properties which repeated: 'the eighth element is a kind of repetition of the first, like an octave in music,' Newlands said. This was the germ of the Periodic Table as it is now known, but the idea didn't get very far.

Dmitri Mendeleef was a professor of chemistry at the university of St Petersburg, and one of the most eminent chemistry lecturers in Europe. In the year 1869, he was composing his magnum opus *The Principles of Chemistry*. On February 17th (March 1st Gregorian, as

Chapter 6. The Moment of Illumination

the Russians were still in Julian time), Mendeleef was scheduled to visit a cheese factory. At breakfast, he jotted down some ideas about the ordering of the sixty-three elements then known. Intrigued, he decided to cancel the visit. He drew up tables, and noticed how the valencies of the elements rose and fell in regular sequences. He made cards for each one, writing the important properties on each card. For several hours, he juggled them about until the arrangement seemed satisfactory. Newlands had forced all the elements into equal rows of seven, but Mendeleef realised that the later periods were longer, and that there were some gaps where elements seemed to be missing. Taking his afternoon nap, he awoke with the idea that he should transpose his table to place the elements in vertical rather than horizontal groups, as he had first arranged them, and thus the Periodic Table was born.

Mendeleef predicted that some elements were still undiscovered and left gaps for them in his table, on which basis he was able to predict the properties of several missing elements. This was highly successful, for three new elements were soon discovered and were similar to Mendeleef's predictions. He published his first table right away, which was just as well because Lother Meyer was coming up with a similar idea. His paper was translated into German and there met with scepticism. Not until the new elements he had predicted appeared was he vindicated. Mendeleef's experience is a classic instance of the eureka insight emerging from a dream-reverie condition.

Ramsay and Helium

1895, 23 MAR 7Q, 2S

'Papa thinks he really has discovered a new gas and Mr Crooks thinks so too.'

Elsa Ramsay, letter of 16 March, 1895.

The discovery of helium by the Scottish chemist William Ramsay is well documented, enabling us to examine this fine eureka experience in some detail. The experience of 'bisociation' as Koestler described it here linked up lines observed in the chromosphere of the Sun and a gas emitted from a rare sample of rock. Having previously discovered argon, Ramsay soon came to hear about a mystery gas emitted by a rare ore from Norway, called 'clevite'. Experts had declared that the gas was nitrogen. Ramsay's letters to his wife, up in Scotland, document the steps of discovery. On March 14th he wrote,

> I have another new gas, I think, from the mineral clevite. There is very little of it, but it isn't nitrogen, and it isn't argon. It has a very distinct but quite different spectrum.

He decided to purify the gas sufficiently so that its spectrum, whatever it was, would be recognisable. He bought up all the clevite available, which wasn't much. On March 22nd, he provisionally called the new gas, 'krypton.' A sample was placed in his spectroscope (at University College, London) together with one of argon so that the two could be compared. He found that:

> There was a magnificent yellow line, brilliantly bright, not coincident with, but very close to the sodium yellow line.

He sent a sample to the physicist William Crookes and on the morning of Saturday, March 23rd, the telegram from Crookes was handed to Ramsay, while he was in his dark room observing the spectra. It stated that the sample had the spectrum of helium, hitherto only observed in the Sun. This 'caused great excitement' in the laboratory. The second lightest element had appeared on earth. Crookes' telegram added, 'Accept my hearty congratulations on a most brilliant discovery,' so he regarded Ramsay as deserving the credit.

In Ramsay's notebook for that historic Saturday morning we find scribbled:

> *Krypton Brilliant yellow <u>not sodium</u>*

then later on:

> *While observing, Crookes telegraphed - Krypton is helium, 587.49. Come & see it. Went & saw it.*

(the number alludes to the wavelength of the gas he had found). On the next day, March 24th, Ramsay explained the excitement in a letter to his wife: 'Helium is the name given to a line in the solar spectrum, known to belong to an element, but that element has hitherto been unknown on the earth. "Krypton" was what I called the gas, knowing its spectrum to point to something new ... It is quite overwhelming, and beats argon'.[12] No-one else was near to reaching this discovery. Sir William Ramsay was awarded the Nobel prize in chemistry in 1904 for his discovery of argon, helium and other rare-earth gases.

Röntgen and X-Rays

1895, 08 NOV   2Q, 7S

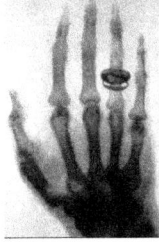

On the midnight hour when Wilhelm Röntgen found X-rays, *seven septiles* were present in the heavens. They were rather close aspects, having a mean orb of only 46 arc minutes. The horoscope for this moment shows these septiles (figure 6.4).

Figure 6.2: X-ray hand

A recent reconstruction of the great E-moment gives its time as well as date:

> The apparition was so awful that Wilhelm Röntgen wondered if he had taken leave of his senses. He could hardly have been more surprised if he had looked into a mirror and no reflection had stared back. He let go of the metal that he had been holding and jumped, startled, by the noise, as it hit the floor. It was approaching midnight on 8 November, 1895.
>
> Earlier that day, he had noticed that whenever he made sparks in the tube, a fluorescent screen at the other end of the laboratory table glowed slightly. This was the signal that he had been looking for, the sign that invisible rays were being produced in the spark tube, crossing the room, and striking the screen, producing the faint glimmer ...
>
> After a late meal Röntgen returned to the laboratory. He moved the piece of lead near to the screen, watch-

ing its shadow sharpen, and it was then that he dropped it in surprise: he had seen the black shape of the metal held by the hand of a dead man. Pulling himself together he slowly opened his clenched fist and looked astonished once again at the dark skeletal pattern of the bones as his hand moved across the face of the screen. Still doubting what he saw he took out some photographic film for a permanent record. Röntgen had made one of the most monumental discoveries in the history of science — X-rays ... If any single moment marks the start of modern physics and science it is that Friday evening of 8 November in 1895.[13]

This appears as a Saturnine E-moment, with lead (Saturn's metal) and the skeleton seen. Saturn was then conjunct the Sun and Mars (5° and 4° orbs), square to Jupiter (2°) and in quintile to Neptune (41′).

That moment constituted the strongest septile aspects over a thirty-day period. The seventh-harmonic harmogram (figure 6.5) shows this. If we zoom in for a close-up of the day of discovery, we see the peak as being at 11 p.m. — i.e. spot-on the very moment given in the above-quoted reconstruction. Two of the septiles were lunar, so they came and went quite quickly, as makes the timing of this event so important. The seven septiles were present for a period of six hours (shown along the base of the harmogram), and it is only the 'power' of these aspects, as takes account of the closeness of orb, which points to the exact time.

Figure 6.3: Wilhelm Röntgen (Roentgen) (*Mary Evans Picture Library*)

Röntgen's discovery of X-rays is often described as a lucky chance: 'a classic case of discovery through accident,' to quote Tom Kuhn.

Chapter 6. The Moment of Illumination

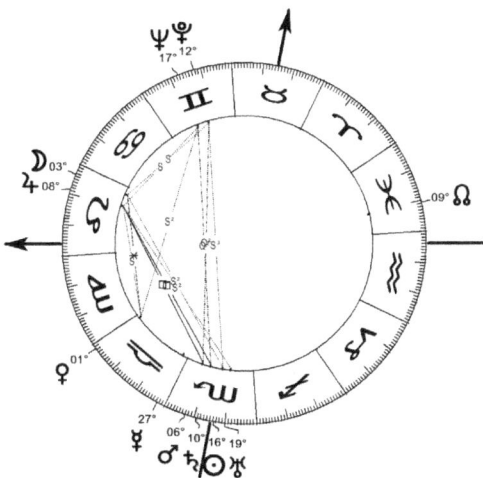

Figure 6.4: Event chart for X-Ray discovery: X-Rays, 08 Nov 1895 23:00hrs, Würzburg, Germany, 49°48'0"N 9°56'0"E (5° orbs used.)

The story, Kuhn explained, showed how what he called 'normal' science gave way to his paradigm-changing 'extraordinary' science: 'He described how Röntgen interrupted a normal investigation of cathode rays because he had noticed that a barium platinocyanide screen at some distance from his shielded apparatus glowed when the discharge was in process. Further investigations — they required seven hectic weeks during which Röntgen rarely left the laboratory — indicated that the cause of the glow came in straight lines from the cathode ray tube.'[14] [15]

When he announced his discovery on December 28th, it was greeted with shock. Lord Kelvin announced that the new rays were an elaborate hoax. Others had to accept the evidence, but were clearly staggered by it. X-rays opened up a strange new world. Within a year after their announcement, dozens of books and pamphlets and over a thousand articles on X-rays appeared. The initial incredulity over X-rays 'gave place upon their confirmation to a delirium of enthusiasm, experimentation, and expectation ... the famous picture of a hand in which the bones thus stood revealed was soon to be found in every city of Europe and America. The realism of this weird picture fascinated all who beheld it.'[16]

Röntgen grew bitter and sank into melancholia. Receiving the

EUREKA

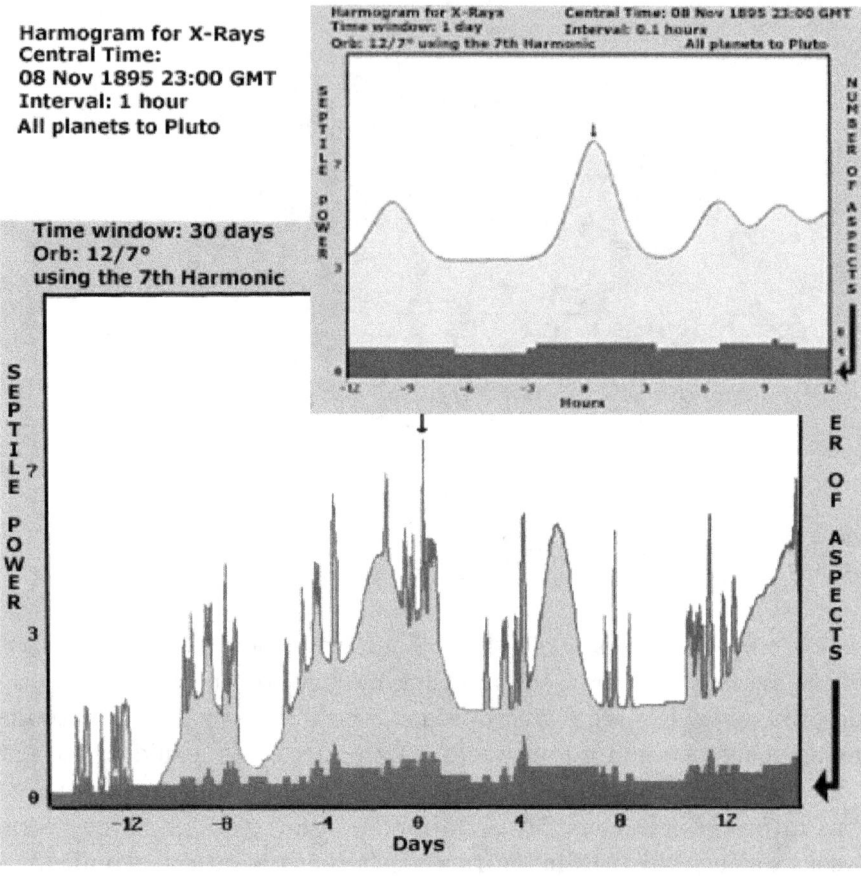

Figure 6.5: 7th-H. Harmogram over thirty days, centred on X-ray E-moment, with close-up of the day centred on 11 p.m.

Nobel Prize in Stockholm, he refused to make any speech or take part in the ceremonies. He died of intestinal cancer, possibly brought on by his exposure to the X-rays, leaving a will ordering the burning of all documents.

Fleming Discovers Penicillin

1928, 03 SEP 4Q, 7S

'Fleming had been waiting for that stroke of luck for fifteen years; and when it came, he was ready for it,' argued Arthur Koestler. Koestler wanted to refute the notion that eureka experiences were

Chapter 6. The Moment of Illumination

brought about by 'mere' chance, as some had argued. Today we know that the discovery of penicillin by Alexander Fleming is far more mysterious than even Koestler suspected.[17] Teams of researchers have tried to ascertain how penicillin could possibly have been discovered in the manner that it was!

Chemistry professor R. M. Roberts, (of the university of Texas), saw Fleming's discovery as a prize instance of serendipity, and added:

> Fleming's life is so full of apparently unrelated events, without any one of which it would not have reached the climax it did, that one 'feels driven to deny their being due to mere chance', as his friend and colleague, Professor C.A. Pannett, said in his eulogy upon Fleming's death.[18]

Fleming adopted the medical profession because his brother was a doctor, went to St Mary's hospital in Paddington, London, where he spent the whole of his life, because he had played against their waterpolo team, and chose bacteriology as his branch of research because someone wanted to keep Fleming, who was an excellent shot, in the St. Mary's rifle club.

Penicillin has been described as the greatest single advance in medicine in the twentieth century. However, the story as told by Sir Alexander Fleming does not reveal all the circumstances, which only came to light much later. One of his colleagues, Dr Ronald Hare, re-investigated the matter, and wrote *The Birth of Penicillin*. During the First World War Fleming was working as part of a team on the bacteriology of war wounds. The antiseptics then available were useless against bacteria, severely damaging exposed tissues and doing more harm than good. One needed a substance that was more lethal to bacteria than it was to tissues, and Fleming was looking for it.

The first step came in 1922, when a drip from Fleming's nose fell onto a dish in his laboratory at St Mary's, and the nasal mucus killed off the baccilli in the culture. Fleming isolated the active agent, which was also present in tears, and called it 'lysozyme'. This was a start, but it was not powerful enough to work as a germ-killer.

Later on, in 1927, Fleming was working with the disease-producing bacteria called staphylococci, growing them on solid gel in round dishes. During August, he went for a vacation. Before leaving, he piled the culture dishes up in the sink and left them to fester. He should have either put them in the fridge or washed them up, but fortunately he did neither. When he returned in September, he was discarding the dishes into a shallow tray containing Lysol, when a colleague happened to walk in and ask how the work was coming on. To demonstrate a point, Fleming took up some of the cultures already discarded but luckily not submerged, and one could see that some were contaminated with moulds. This was hardly surprising after leaving them over the summer.

One culture in particular caught Fleming's attention; after looking at it for a while he said, 'That's funny.' The colonies of bacteria had been 'lysed', i.e. they had grown transparent owing to the bacteria breaking down. The two of them agreed that the situation was very odd, and Fleming promptly made a subculture of the mould. In the tea-break at eleven o'clock in the morning he took it upstairs to show to his colleagues, who were unimpressed.

It was for long assumed that Fleming had merely had the perspicacity to notice what others had seen and ignored. The plot thickened, however, when it was realised that ordinary penicillin does not 'lyse' staphylococcal colonies, but prevents them from developing, and if added afterwards has no apparent effect. To quote from Beveridge's *Seeds of Discovery:*

> Hare [a biochemist] found that it was very difficult indeed to produce lysis of staphylococcal colonies, even when the cultures were deliberately inoculated with a known penicillin-producing mould. Only after a long series of experiments was he able to discover the very special conditions required to produce the phenomenon. He showed that the rare oddity that Fleming observed could only have arisen as a result of a chain of events that produced just the right, unusual conditions. Fleming must have inoculated a culture dish which at the same time became accidentally contaminated with a mould spore, and instead of putting it in the incubator as is normal practice, it seems he must have got it mixed up with old cultures and inadvertently left it on the bench. The dish lay there during Fleming's vacation and it so happened that there was an unseasonable cold snap in the weather that pro-

Chapter 6. The Moment of Illumination

vided the particular temperature required for the mould and the staphylococci to grow slowly and produce lysis of the staphylococci colonies near the mould colony. Hare remarks that the odds against this combination of unlikely events happening just by accident, which of course it did, must be astronomical.[19]

Later Fleming would say, 'There are thousands of different moulds and there are thousands of different bacteria, and that chance put the mould in the right spot at the right time was like winning the Irish sweep.'

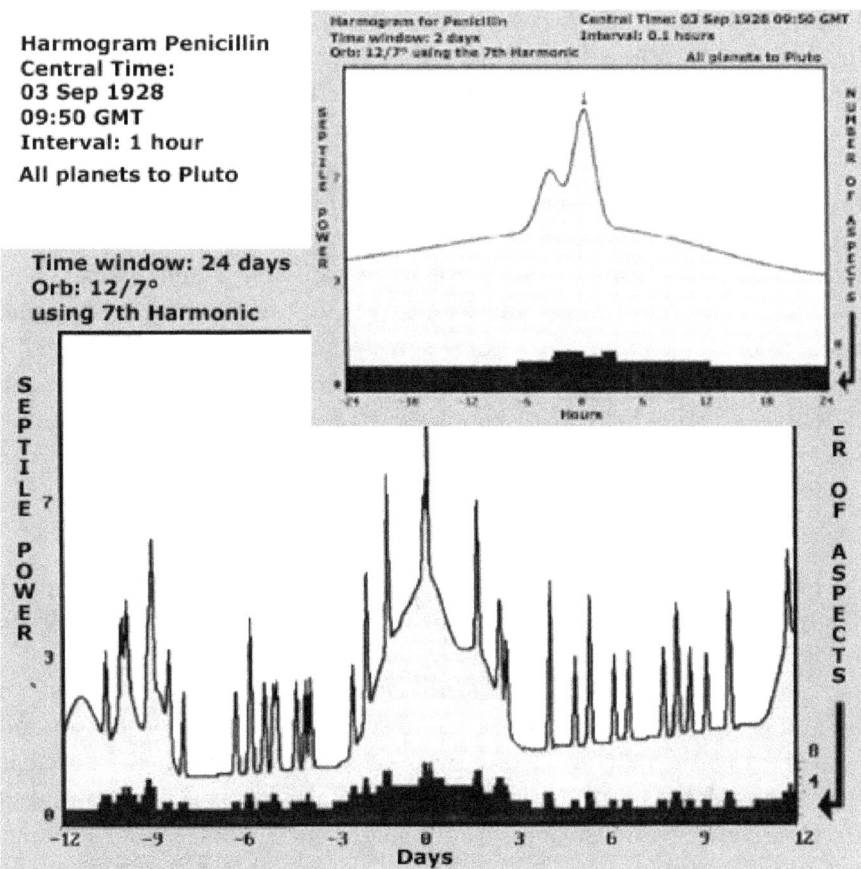

Figure 6.6: Eureka moment for Penicillin, 7th Harmonic, 24-day period, with close-up of a 2-day period.

101

Millions of pounds have been spent researching for antibiotics since that time, yet penicillin remains unsurpassed. It happens that most strains of the mould 'Penicillium' do not produce penicillin. A world-wide search of strains of Penicillium capable of breeding high yields of this antibiotic showed that Fleming's spore was one of the three best producers. Experts now believe that the spore probably drifted up the stairway from a laboratory where research on moulds was proceeding, and not in through the window as legend has it. Fleming tried to produce penicillin in a form usable for treating infections, then finally gave up. Neither he, nor anyone else at the time, imagined the potential of penicillin as a therapeutic agent, and for nine years nothing further happened. One Jupiter-cycle later, Fleming saw his great discovery become of practical utility (described in chapter 9: When Inventions Worked), developed by Florey and Chain at Oxford, then used in World War II.

'The story of penicillin has a certain romance in it,' remarked Fleming, 'and helps to illustrate the amount of chance, or fortune, of fate, or destiny, call it what you will, in anybody's career.' The harmograms (figure 6.6) show the powerful septile peak at the moment when Fleming noticed that bacteria colonies on a jellied dish, had grown transparent. Mars and Venus were in an exact $(2')$ biseptile, when the Moon came into exact septile and triseptile aspects with these two bodies at 10.30 a.m. For the timing of the moment, we are indebted to Mike Edwards, who watched a BBC Horizon programme which reconstructed the sequence of events over the discovery, and informed us of the matter.

Tombaugh Finds Pluto
1930, 18 FEB   3Q, 1S

Sir Percival Lowell was an authority on the canals of Mars which he had seen and mapped from his observatory. He wrote books and lectured on the matter. His Lowell Observatory stood in the clear desert air of Flagstaff, Arizona. The canals, he explained, pointed to an advanced civilisation on our neighbour planet Mars, a planet which had learnt to live at peace with itself as shown by the way the canals extended all over the Red Planet. When, in the early years of this century, Lowell's case dissolved amidst ridicule and derision,

Chapter 6. The Moment of Illumination

he sought for a means of restoring the shattered credibility of his observatory. He determined to locate a new planet.

Percival Lowell spent eleven long and futile years searching for a planet beyond Neptune. He believed that Uranus and Neptune had slight discrepancies in their orbits, which indicated the position of 'planet X'. Long computations enabled Lowell to determine the characteristics of this outermost planet — its inclination, its perihelion, its distance from the Sun, and which zodiac signs it should be in. He took photographs at several day intervals of the same star-fields. Any point that had moved would either be a comet, an asteroid — or, the new planet. Exhausted by the struggle, Lowell died at the comparatively young age of 61. To quote Asimov, 'Lowell's long failure had taken the heart out of the search for most people'.[20]

Lowell left a trust fund to be used in the search for the transNeptunian planet. A new telescope was built at Flagstaff from it, able to photograph the stars with a long exposure. The new director of the observatory received a letter from a keen young amateur, describing the canals of Mars in great detail. He had seen them with his homemade telescope. That was a fine recommendation, as Lowell had more than anyone else put the phantom canals of Mars on the map. The young man, whose name was Clyde Tombaugh, was recruited. Tombaugh's chart had his Mars conjunct that of Lowell within one degree, i.e. they were both born with Mars in the same position.

Tombaugh developed a 'blink' technique, whereby images the same area taken a day or so apart could be rapidly alternated, causing any moving object such as a comet to move between its two positions, while all the rest of the star-fields remained motionless. Each photograph taken contained thousands of stars. Tombaugh began in the Fall of 1929, and by February of the next year his quest was fulfilled, amidst the stars of Gemini.

At four o'clock in the afternoon, he became convinced that he had seen it:

> I was walking on the ceiling. I was now 100% sure.

At the great moment, Tombaugh recalled how:

> A terrific thrill came over me. I switched the shutter back and forth, studying the images. Oh! I had better look at

> my watch and note the time. This would be a historic discovery. Estimating my delay at about three minutes, it would place the moment of discovery very close to four o'clock. For the next forty-five minutes or so, I was in the most excited state of mind in my life.[21]

He rummaged through other photographic plates, to see if they showed an object moving at the required speed to be beyond Neptune:

> I can hardly describe to you how intense was the thrill I felt. I was looking through the hand magnifier identifying the configuration of the stars. I could hardly see them; my hand was shaking with excitement.

He strolled over to the manager's office, trying to stay calm, and said, 'I have found your planet X.' Later that day, his mother was hanging out the washing when Clyde's father told her the news, 'Clyde's discovered a planet!' Two days later, Pluto was first viewed by human eye. On March 13th it was announced, and the press from all the world swarmed up Mars Hill to the Observatory.

Scientists were awed by the precision of Lowell's predictions for 'planet X'. To quote from the Scientific American, 'The Orbit, now that we know it, is found to be so similar to that which Lowell predicted from his calculations fifteen years ago, that it is quite incredible that the agreement can be due to accident.'[22] Lowell had predicted 205° perihelion, and the true value was 212°; had predicted 10° inclination to the ecliptic — far larger than any of the other outer planets — and the actual was 17°; and had predicted a zodiac position less than 5° away from the true position when found!'[23]

Was this not another awesome indication of Science's ability to predict the unknown? A problem gradually arose in the course of time, as Pluto's perceived size kept shrinking, until eventually it became clear that it was far too small to have any detectable effect in perturbing the outer planets ... All Lowell's labyrinth of computations had been just as ghostly as his canals of Mars. The sole difference was, that in the latter case he had been wrong, while in the former, he was right. Pluto had been in the sole part of its orbit which was close to Lowell's calculated position. It crossed the ecliptic in September of 1930. At any other position, Tombaugh could not have found it.

Chapter 6. The Moment of Illumination

Pluto was less than one degree away from its ascending node when discovered. Nodes are traditionally viewed as power-points, as might be especially appropriate for Pluto because its orbit is so eccentric, making the arrival at its node a fairly distinct event.

His transits at the discovery were:

Table 6.2: Natal: Tombaugh Pluto Discovery Orb

Mercury	conjunct	Mercury	0°26'
Uranus	square	Uranus	2°00'
Saturn	conjunct	Uranus	2°00'
Venus	conjunct	Saturn	0°24'

The exact Mercury return is appropriate for this thrilling moment, while the far slower Uranus square return reminds us of Clyde Tombaugh's youth, at twenty-six years of age. Usually this return is completed a year or two earlier, but it can vary. The two Saturn transits remind us of the event's profound scientific meaning.

But, was it an E-moment? Initially, when commencing the eureka project, it seemed that notable astronomical moments, such as the discovery of Pluto or Galileo spotting the moons of Jupiter, were ideal material, but they have turned out to be problematic. They were included and scored early on, so we did not feel at liberty to remove any of them; but, if there is one that should be removed, it is surely this one. Tombaugh was just grinding mechanically through a semi-automated photographic procedure, knowing in advance what he was looking for, which is in a sense the antithesis of a eureka moment.

Our list of 21 E-moments was sent to Professor Grattan-Guinness, an editor of the journal *History of Science* and also of a journal on the history of logic, who kindly agreed to comment upon it. He seemed to find the list of E-moments acceptable, with one exception. He said, 'I'm not struck by case 16, as Tombaugh will only have assessed his data then: no special intellectual insight. He had actually photographed Pluto 9 months earlier, but did not realise.' (Personal communication, quoted with kind permission.) The (Q+S) score of this event is below average, so perhaps the cosmos is telling us something here.

As regards Lowell's ability to make such a prediction — and, indeed being the only mortal man to have his initials used for a planetary glyph — there was an exact alignment of Lowell's chart with the new sphere: his natal Moon at 17° 43' Capricorn was conjunct his MC and opposite to Pluto's position at its discovery (17° 46' Cancer) within minutes of arc.

Meitner and Nuclear Fission
1938, 24 DEC   1Q, 4S

What appears to be the only female eureka scientist played a key role in smuggling the secret of atomic fission out of Nazi Germany. Today there is a Hahn-Meitner Institute in Berlin marking where Otto Hahn and Lise Meitner worked in the 1930s on radiochemistry. 'Chance ... gave the Americans rather than the Germans the harvest of nuclear fission discovered by Otto Hahn in Berlin.'[24] Chance, or whatever one calls it, worked through one woman.

Meitner had been a valued co-worker of Hahn for thirty years. But, having Jewish ancestors, she had to flee from Nazi Germany. She escaped, then friends secured her a place outside Stockholm with the Swedish Academy of Sciences. 'She travelled to that far northern exile, to a country where she had neither the language nor many friends, as if to prison.'[25]

Fission had been taking place ever since Fermi started his element transmutation experiments in the early 1930s, but was not recognised as such. Otto Hahn came near to realising what was happening in his uranium bombardment experiments. In December 1938, he obtained results which were 'at variance with all previous experience', but he could not interpret them. Barium was somehow appearing from uranium in his irradiation experiments. Baffled, he wrote to Lise Meitner about this top-secret matter, saying 'Perhaps you can suggest some fantastic explanation. We understand that it *can't* break up into barium ... ' He closed by wishing her a 'somewhat bearable' Christmas.

'Your radium results are very amazing' she wrote by return, on December 21st. Some friends living in the small village of Kungalv near Stockholm considerately invited the two physicists Otto Frisch and his aunt Lise Meitner to dinner over Christmas, as they were both alone. Otto Frisch, a co-worker of the great Danish physicist Niels

Bohr in Copenhagen, had been studying the behaviour of neutrons, and was hoping to discuss this with his aunt, but she had something quite different on her mind. Over breakfast she perused Hahn's letter of December 19th, and asked Frisch to read it. He glanced over it and said 'Barium, I don't believe it. There's some mistake.'

As Otto Frisch recalled: 'There, in a small hotel in Kungalv near Goteborg I found her at breakfast brooding over a letter from Hahn. I was sceptical about the contents — that barium was formed from uranium by neutrons — but she kept on with it. We walked up and down in the snow.'[26] They walked towards the market place of Kungalv, over the frozen river and into the woods beyond, then sat down on a log. Frisch was protesting that the notion of a nucleus splitting was impossible. They then realised that the uranium nucleus was behaving like a liquid drop, which 'might elongate and divide itself.' They performed energy calculations, using the very new $E = mc^2$ equation, and it all seemed to fit. According to C. P. Snow, (in *The Physicists*) the couple had been out in the snow for 3 hours when it came to them, and they went out after breakfast.

At first, they were wary of telling anyone because it seemed so outrageous. Frisch told Neils Bohr about their theory when he returned to Copenhagen, whereupon the latter struck his forehead and said, 'Of course, that's the way it must be! What fools we've all been!' Frisch decided to call the idea, 'fission.' Bohr was on his way to Washington for a conference, and took the story with him. No historical account permitted inference as to the date of this Christmas event, until the publication in 1986 of Rhodes' *The Making of the Atomic Bomb* where it was located as being on the morning of Christmas Eve.

Townes and the Laser
1951, 06 APR   3Q, 7S

> 'There is a tremendous emotional experience (in scientific discovery) which I think is similar to what some people would normally describe as a religious experience, a revelation.'
>
> Charles H. Townes[27]

Recollected professor Charles Townes,

> The laser was born early one beautiful spring morning on a park bench in Washington, D.C. As I sat in Franklin Square, musing and admiring the azaleas, an idea came to me for a practical way to obtain a very pure form of electromagnetic waves from molecules.[28]

His account describes one of the very few documented, scientific E-moments on record within the United States. Regrettably, this first account appeared more than three decades after the event, but alas this is not unusual. It gives no date. A letter was sent to Professor Townes, and he sent in reply a photocopy of his laboratory notebook which gave the key date, of April 26th.

Figure 6.7 shows 'septile power' over a period of twenty-four days. It climbs to its peak value at the very historic instant when Charles Townes described having his insight as to how a laser could work, in April of 1951. I received a letter from Townes, recalling the very time that morning when the idea dawned upon him, describing how he had then been chairing a conference on short-wave radiation in Washington DC:

> I was in the habit of waking up early, and I woke up at perhaps 6:30 in the morning. It was a bright sunny morning, and I left the room in order not to disturb Schawlow and to sit in the nearby park to think over the day's tasks. While puzzling over how we might make progress towards shorter waves, and why we had so far failed to make a great deal of progress, the idea of the maser came to me as the only practical method I could visualise for getting quite short waves. That was still before breakfast, and must therefore have been before about 7:30 a.m. I remember particularly the beautiful azaleas as I sat on the park bench ...

He discussed his new idea over lunch with his students at the University of Columbia. (N.B. What Townes here alluded to as the 'maser' was the precursor of the laser.) This letter makes Townes' E-moment the best timed of all our E-moments.

Chapter 6. The Moment of Illumination

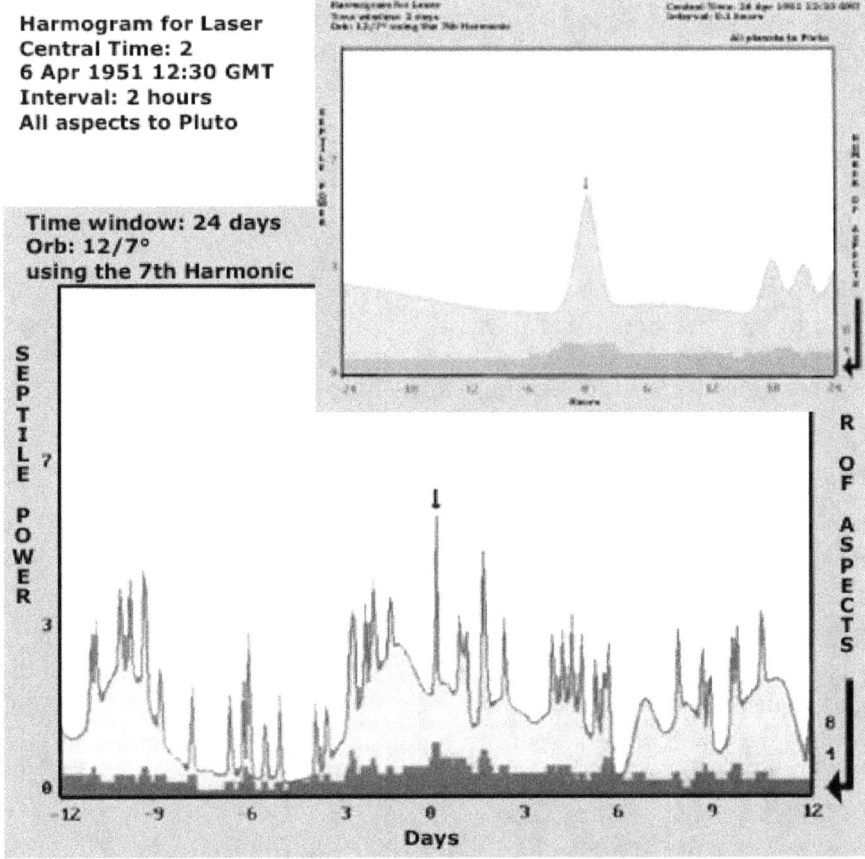

Figure 6.7: Eureka Moment for Laser, 7th harmonic, 24-day period, with close-up, 07:30hrs EST (12:30hrs GMT), 26 Apr 1951 Washington DC

Septile power over a 24-day period is shown in figure 6.7, peaking at the moment when Charles Townes was reclining on his park bench admiring the azaleas. How closely did the ebb and flow of septile power define his historic moment of inspiration? To answer that, we zoom into a close-up of ±1 day around the event, as shown, which peaks at 7:30 a.m., from which it can be seen that the inspiration came to Townes at the very optimal time of the day and month with regard to the septile aspects present.

The sharp peaks shown on the harmogram are lunar septiles, which come and go quickly, and at 7:30 a.m. (12:30hrs GMT) there was a double lunar septile: Saturn and Venus were in a close (13′)

109

biseptile, and the Moon came into exact aspects with the two — a biseptile with Saturn, and a triseptile with Venus. Surely this pattern shows the very architecture of inspiration.

Fermat's Last Theorem[29] — Solved
1932, 22 JUN   3Q, 3S

Probably in the year 1630, Pierre de Fermat had scribbled in the margin of a mathematical treatise,

> I have discovered a truly remarkable proof which this margin is too small to contain.

— the alleged proof being that no equation of the form $x^n + y^n = z^n$ could exist, provided that the solutions are non-zero, the numbers are integers and $n > 2$. Fermat was a lawyer who only ever published one mathematical paper, anonymously. His son collected his writings, posthumously, and thus this remark was preserved. Mathematicians surmise that actually Fermat had not found any such general proof, and realised that. He was, however, able to solve the special cases of $n = 3$ and $n = 4$, and he sent out challenges to other mathematicians. Through centuries of struggle, mathematicians failed and failed again to find any such general proof.

Over a seven-year period, at Princeton University, the British mathematician Andrew Wiles worked in secrecy, with no-one except his wife, Nada, knowing of his secret obsession. He did not want people gossiping about his aim, the fabled first prize in mathematics, as this would have put him under too much pressure. Alone, he struggled with the Taniyama-Shimura conjecture. A decade earlier a Japanese mathematician had claimed to have found a solution, but this fell apart under examination. On a New York subway someone had written

$$x^n + y^n = z^n \text{ no solutions}$$
I have derived a truly remarkable proof of this,
but I can't write it now because my train is coming.

Wiles was invited to give a lecture about number-theory at the Isaac Newton Institute at Cambridge, and rumours started to spread.

Chapter 6. The Moment of Illumination

On the 22nd June, 1993, e-mails hinted that his lecture the next day would culminate in a solution to Fermat's Last theorem! Wiles covered several blackboards with a dense mixture of Greek symbols and algebra, then after half an hour, at half-past ten in the morning he concluded, 'And this proves Fermat's Last theorem.' Two hundred mathematicians burst out in cheering and applause. Even those who had anticipated the result grinned in disbelief. One person commented,

> 'I've never seen such a glorious lecture. Full of such wonderful ideas, with such dramatic tension, and what a build-up. There was only one possible punch-line.'

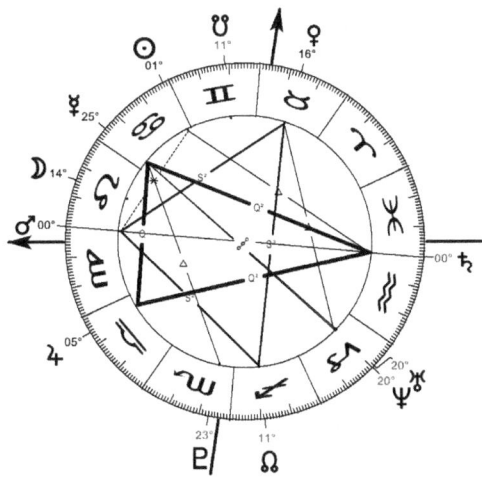

Figure 6.8: Event Chart for Fermat's Last Theorem Solved E-moment highlighting quintile aspects, 10:30hrs BST (09:30hrs GMT) 23 Jun 1993 Cambridge, UK. (Addey orbs used.)

The next morning saw that eureka cry echoing around the world's newspaper headlines, for the world's most famous mathematical problem had been solved.

'At last, Shout of "Eureka!" in *Age-Old Mathematical Enquiry*', proclaimed the *New York Times*; 'The Number's Up for Maths last Riddle' was *The Guardian* headline. That lecture was the first record of such a solution, and was a public event. 'Overnight Wiles became the most famous, in fact the only famous, mathematician in the world.' Professor Shimura was pleased to hear about his conjecture being vindicated, even though his colleague Taniyama had committed suicide some years earlier. Some people felt that the establishing of that conjecture was the most important part of the lecture, because of its important applications. A clothing firm asked Wiles if he would endorse their range of menswear.

The theorem is essentially triangular, as Pythagoras' theorem $a^2+b^2 = c^2$ applies to right-angled triangles. There are whole-number solutions for a, b and c, while Fermat's Theorem affirmed that this could not apply in higher dimensions, using a^3, etc. The solar system was then forming *triangle structures* woven of trines, sextiles and oppositions — as well as quintile and septile aspects. The quintile chart of this moment (figure 6.8) gives us a lovely 'divine triangle' as mathematicians call it (there is no astrological term for this shape, made of quintiles and biquintiles), with tight orbs, *pointing at Saturn*, right on the horizon during the lecture's half-hour.

Turning to the septiles in this discovery chart, there are five if the lunar nodes are included. Again we see a remarkable triangle, symmetrical and composed of bi- and tri-septiles, which expresses the slope angles of the Great Pyramid. (See page 55, chapter 4: The Angles Of Inspiration). Thus two rather appropriate mathematical forms appeared in the heavens at this mathematical e-moment. The proof was a gigantic argument, intricately constructed, and no wonder he needed some help from ... Mercury. He is now Sir Andrew Wiles, the only mathematician ever knighted for his work.

Lost E-moments

A few members of our eureka group have 'lost' E-moments, which means that we were unable to find their dates. As discussed earlier, such moments are often mythical, i.e. dreamed up years later; however some of them — Nicola Tesla and Count Louis De Broglie — do seem to be genuine.

There is a dramatic E-moment for Pasteur often related, though it has recently been dismissed as quite mythical. When Pasteur discovered crystal optic rotation, he is said to have 'rushed out of the laboratory, not unlike Archimedes ... ' in the Paris School of Medicine, and embraced the nearest chemical assistant in the hall, saying 'I

Chapter 6. The Moment of Illumination

have just made a great discovery ... I am so happy that I am shaking all over and am unable to set my eyes again to the polarimeter!' This was supposed to be in April 1848. An article in the history of science journal *Isis* has reviewed the documents and concluded that the event was mythical, recollected in later years to give a dramatic flourish to his discovery.[30]

In 1882, Nicola Tesla had been told that an alternating current electric motor was an impossibility. He felt certain that it had to be feasible, but was not sure how, and became ill due to this mental struggle. One day he was walking with a friend in Belgrade park, and as the Sun set, he was reciting some lines from Goethe's 'Faust', when:

> As I uttered these inspiring words the idea came like a flash of lightning ... I drew with a stick in the sand the diagrams shown six years later in my address before the American Institute of Electrical Engineers, and my companion understood them perfectly. The images I saw were wonderfully clear and sharp and had the solidity of metal and stone, so that 'for many years afterwards my life was little short of continuous rapture.'[31]

We were advised by a student in the Muzej Nicole Tesle in Belgrade that the answer as to when Tesla reached his rotating magnetic field concept may be 'somewhere among 70,000 documents of Tesla's personal correspondence which is not yet explored' but for now it must be accounted lost; as too are his great discovery moments.

Count Louis De Broglie had an almost-dateable E-moment, shortly before his September 10th paper on the subject. He described how, 'After long reflection in solitude and meditation, I suddenly had an idea ... ' and that idea was 'his epochal new principle that particle-wave duality should apply not only to radiation but also to matter.'[32]

113

References

1. Richard Rhodes, *The Making of the Atomic Bomb*, 1986, p.1.

2. T. H. Levere, *Poetry realised in Nature, Samuel Taylor Coleridge and nineteenth Century Science*, CUP 1981, p.72.

3. Max Caspar, *Kepler*, 1959, 1993, Dover, New York, p.267.

4. H. Lewis McKinney, *Wallace and Natural Selection*, 1972, p.139.

5. Augustine Brannigan, *The Social Basis of Scientific Discoveries*, 1981, CUP, p.28.

6. Koestler, *AOC* p.140.

7. W. Beveridge, *Seeds of Discovery*, 1965, 1980, p.68.

8. H. E. Gruber, 'Aha Experiences', *History of Science*, 1981, 19, p.43.

9. H. L. McKinney, *Wallace and Natural Selection*, Yale 1972, p.135; History of Science, Vol. 44, 2006, pp.1–28.

10. A. R. Wallace, *My Life*, 1908.

11. N.K., http://www.dioi.org/kn/neptunestory.pdf; N.K., 'A Hiatus in History, the British claim to Neptune's Coprediction', *History of Science*, 2006.

12. Morris W. Travers, *A Life of Sir William Ramsay*, 1956, p.145.

13. Frank Close, *Lucifer's Legacy, the Meaning of Asymmetry*, 2000, pp.77,79.

14. T. S. Kuhn, *Structure of Scientific Revolutions*, 1967, 1970, p.57.

15. D. K. Claxton, *Wilhelm Röntgen*, 1970, p.144.

16. E. V. Heyn, *The Fire of Genius, Inventors of the Past Century*, 1976, p.278.

17. Koestler, *AOC* p.194; Beveridge, 1980, pp.20–24.

18. R. M. Roberts, *Serendipity*, 1989.

19. W. Beveridge, op. cit. (7), p.23.

20. Isaac Asimov, *Quasar, Quasar, burning Bright*, NY 1978, p.170.

21. Prof. H. N. Russell of Princeton, 'More about Pluto,' *Scientific American* Dec. 1930, pp.446–7. Lowell predicted that Pluto would reach its perihelion in 1991, which it did in 1989. 'See also N.K., Pluto R.I.P., 1930–2006?' http://www.astrozero.co.uk/articles/plutorip.htm.

22. A. Lowell, *A Biography of Percival Lowell*, 1935, p.199.

23. Clyde Tombaugh and Patrick Moore, *Out of the Darkness, The Planet Pluto*, 1980, p.127; David H. Levy, *Clyde Tombaugh, Discoverer of the Planet Pluto* AZ 1991, p.5. For discussion of Clyde Tombaugh's 'terrific thrill', see Brian Taylor,

References

'The discovery of Pluto: an unbidden omen' in *Orpheus, Voices in Contemporary Astrology* Ed. Suzi Harvey, 2000, pp.247–330.

24. R. W. Clark, *The Scientific Breakthrough*, 1974, p.10.

25. Rhodes, op. cit. (1), p.236.

26. Otto Frisch & John Wheeler, 'The Discovery of Fission', *Physics Today*, November 1967, pp.43–49,47.

27. Berland, T. *The Scientific Life*, MIT 1962 (quoted in Current Biographies, 1963). Townes confirmed the validity of this quotation. Townes' *How the Laser Happened, Adventures of a Scientist* 1999 touches on the topic (pp.55–9).

28. Charles Townes, *Science*, November 1984, p.153; R. Y. Chiao, *Amazing Light: a volume dedicated to Charles Hard Townes on his 80th Birthday* Springer 1996, p.5 shows his notebook page with date of 26th April; see also I. Flatow 1992, pp.119–121.

29. Barry Mazer, *Fermat's Last Theorem*, 2002, p270, 272. http://www-gap.dcs.st-and.ac.uk/~history/HistTopics/Fermat's_last_theorem.html.

30. 'The Case of Optical Isomerism' Geison and Secord, *Isis*, 1988, 79, pp.6–36. On the other hand, Prof. Roberts in 'Serendipity' (19) spent several pages describing this E-moment without questioning its authenticity.

31. Marc Seifer, *Wizard the Life and Times of Nikola Tesla*, 1998, p.22.

32. A. Pais, *Subtle is the Lord, the Science & Life of Albert Einstein*, 1982, p.436.

CHAPTER 7

Silhouette of the Septile

'Those who deny the influence of the planets violate its clear proofs, which it does not become people of sound judgement to contradict'

Tycho Brahe[1]

QUINTILE AND SEPTILE ASPECTS TURNED up in the birth charts of eureka-scientists as well as in their E-moments, with the latter aspect being more strongly present than the former. We now take a more general look at just what aspects are involved, and at their structure. John Addey said that a 'series' of quintile and septile aspects could be involved.[2] What would that signify? We start by performing a 'harmonic analysis' on the data-sets.

The Harmonic Pattern

The concept of 'harmonic' alludes to the group of aspects formed by dividing up the circle of the zodiac by a number, so that the number expresses the harmonic. A harmonic includes all such angles on the ecliptic circle except for the conjunction (as represents an angle of zero degrees). We saw earlier how the sixth 'harmonic' includes the two trines (120°) and two sextiles (60°) plus one opposition (180°); where John Addey's orb principle gave the 6th harmonic an orb of $(12 \div 6) = 2°$.

We followed a comparable procedure in setting up the harmonic analysis shown on the natal-group data. Table 7.1 displays the total number of aspects within the natal Eureka-group (with Addey orbs), together with expected frequencies (from computer sampling through the centuries, clocking up scores from times chosen randomly, see appendix B: Statistical Procedure), and also gives some percentage excesses or deficits.[3]

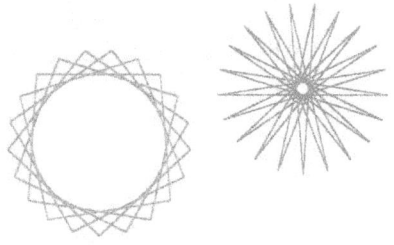

Figure 7.1: Septile 'overtones', angles of the 42nd harmonic, (a 21-fold division of the circle).

The more conventional harmonics such as the 4th and 6th (square and sextile aspects) are in deficit, as also is the 12th, which contains all the traditional Ptolemaic aspects. Other prime-number aspects such as trine, 11th or 13th, show little. Harmonics 2 (*not shown*), 4, 6 and 12 are in deficit. These pertain to stability of structure — for example, use of the hexagon lattice in a beehive — while the achievement of these eureka scientists lay in breaking away from established categories of thought, where the mould-breaking influence of the prime numbers 5 and 7 was required. The traditional aspects are all embedded within the fabric of the zodiac, as multiples of 30°, but these aren't.

A similar harmonic analysis was performed for the group of E-moments. This showed no remarkable excess or lack of aspects except within the septile series, which were as follows:

This table is telling us something about the special character of the sky when the big eureka insights happened. It shows wonderfully how a strong seventh harmonic has reverberated up into its higher frequencies or 'overtones,' as fully endorses John Addey's theory, and also validates in a general sense the idea that music theory is relevant to how celestial aspects work. No-one has ever taken seriously Addey's proposals concerning micro-aspects (made chiefly within the context of slow-moving outer planets), and they contradict Keplerian guidelines for permitted aspects. For this reason, one would

---

*Table 7.1: For harmonic expected frequencies, see graph B.1, appx. B.

†Table 7.2 uses Addey orbs in the form: $(12 \div n)°$.

Table 7.1: Natal Group aspect frequencies.*

Harmonic	Aspect	Observed	Expected	Excess
3	trine	29	27.0	
4	square	27	32.0	−15%
5	quintile	43	35.0	+22% ⇐
6	sextile	30	37.0	−20%
7	septile	55	38.0	+42% ⇐
8	octile	44	40.0	
9	novile	32	40.8	−21% ⇐
10	decile	40	41.0	
11	undecile	49	41.0	+17%
12	dodecile	35	42.0	−17%
13		38	42.0	−13%
14		40	43.0	
15		53	44.0	+19%

Table 7.2: The 7th harmonic In The E-Moment Group using Addey orbs.†

Harmonic	Observed	Expected	Excess
7	65	55.4	+52%
14	75	56.0	+34%
21	86	56.5	+52%
28	82	57.0	+44%

like to see these results re-analysed using a more accurate program — the high-frequency 'overtone' harmonics use very small orbs.* If their presence was confirmed, this would mean that the harmogram programs of the future — used to find dates for musicians or philosophers, seeking an inspirational time to meet — would include these septile 'overtones', as shouldn't be too difficult.

---

*Nowadays, home computers have accurate programs for planetary positions, as can find planetary longitudes within seconds of arc. However, it isn't easy to access these most accurate programs for constructing data-analysis programs. Accordingly, a higher resolution program was used (by Kevin Hawley, Manchester) accurate to a fraction of a minute in historical time. This turned out to *enhance* the levels of significance, for both the natal and event groups, as a result of which slight discrepancies arise: e.g., table 7.1 gives 40 quintile aspects, while the true score (i.e. as more exactly computed) was 41.

Planets of Inspiration

Which planets help with scientific inspiration? To answer this, we first look at planetary frequencies from summing all the quintile and septile aspects within the natal eureka group. Comparing these scores with those astronomically-expected gave the results seen in table 7.3.

Table 7.3: Planetary frequencies in Natal Group — quintile plus septile aspects.

Planet	Observed	Expected	Excess
Moon	27	16.0	+68%
Pluto	26	15.9	+63%
Uranus	24	15.9	+50%
Jupiter	22	15.8	+39%
Venus	17	12.8	+32%
Saturn	20	15.8	+26%
Neptune	19	15.9	+19%
Mars	18	15.3	+17%
Mercury	12	12.8	−06%
Sun	11	12.0[†]	−08%

We are startled to see the Moon at the top and the Sun at the bottom of this table; though one should be wary of concluding too much from this small sample alone. It seems a rather alchemical result — would these lunar aspects be telling us that the scientists are, by nature, reflective? Uranus and Pluto have the highest frequencies of any planet — astrologers consulted were virtually unanimous that Uranus was the most pertinent planet. The same analysis was performed for the eureka-moment group and it gave the results seen in table 7.4.

This is a quite different distribution from that of the natal-group. It has a peak excess of solar aspects — the shining beams of the Sun have irradiated these eureka-moments. Saturn comes second with a remarkable 70–80% excess of these creative aspects, a planet highly appropriate for scientific discoveries. Neptune the 'divine dreamer' comes third, likewise expressing the character of these illuminative

---

[†]Table 7.3: There is a lower chance-expectancy value for the Sun, Mercury and Venus, since these remain close to each other and so cannot form certain aspects.

Chapter 7. Silhouette of the Septile

Table 7.4: Planetary frequencies in Event (E-Moment) Group — quintile plus septile aspects.

Planet	Observed	Expected	Excess
Sun	33	17.3	+90%
Saturn	40	22.7	+76%
Neptune	34	22.9	+48%
Pluto	34	22.9	+48%
Moon	33	23.0	+43%
Uranus	32	22.9	+39%
Venus	24	18.4	+30%
Mercury	17	18.4	−07%
Mars	19	21.9	−13%
Jupiter	19	22.7	−16%

moments. Nimble Mercury has played no part in these grand moments of illumination, while Mars and Jupiter are positively in detriment, and have not helped at all at these times of scientific insight.

Graphs of Celestial Influence

We are now in a position to visualise the very structure of the septile, as it has functioned within the historic data, at moments of illumination. In the past, sages have pondered the question, concerning the shape of a celestial aspect, while persons of shallower intellect have dismissed the matter. Arm-waving conjecture had seemed all that one could ever hope for. We here give a definite answer, the first.

First let us see how the strength of these aspects varied with orb, which tells us how close is the angle between the two spheres to the geometrically-exact value for the aspect. We use the septile aspects, as replicated so well through the two eureka groups.

Figure 7.2 (a) shows the septile 'aspect profile' for the group of birth charts of Eureka scientists. This tots up all the septiles and breaks them down by orb, i.e. how close they were. The 'Addey orb' we used is shaded, and the chance-level is shown by an horizontal line. It shows how the main excess of septiles lay within a degree of orb, though the 'Addey orb' hasn't done badly. In her book, *Aspects and Personality*, (1990) the eminent astrologer Karen Hamaker-Zondag

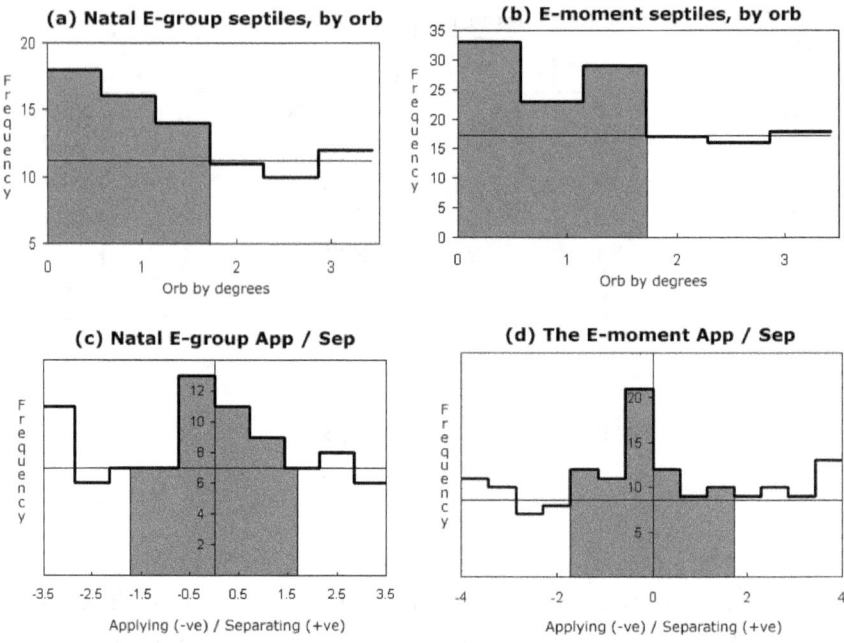

Figure 7.2: The Orb of a septile: (a) Natal Eureka group, and (b) the Eureka moments. Figures (a) & (b) show the orbs 1–2° along the base. For (a) the 'Addey-orb' has been split into three, where the expected value per bar is 11, thrice which is 33, the total 'expected' value. The score within the Addey orb is $18 + 16 + 14$ or 48, and turning back to table 7.1 we find a comparable value. In figures (c) & (d), the aspects have been split into applying (−) and separating (+).

recommends an orb of one degree for the septiles. Figure 7.2 (b) shows the group of E-moments treated in the same way, for the orb of their septile aspects, and this more clearly shows that a smaller orb would have been better. Having the main excess at a tight orb is a good sign as indicating that the effect is no mere artifact.

The notion of orb doesn't distinguish between *applying* and *separating* aspects. To give an example, suppose the Moon was 95° away from it, then it is coming up to a square aspect; then later, when only 85° away, it would have the same orb but be in a 'separating' aspect, i.e. it is moving away from the square. The *time-structure of a celestial aspect* is indicated by figures 7.2 (c) and (d). These distinguish between the aspects which were about to happen, coming up to the moment of exactitude, and those which had already taken place so that they were separating. The computer distinguished between these two conditions, by comparing the speeds of each pair of planets that were in septile aspect. (The situation is complicated by planets going retrograde, but M.O.'s program coped with this.)

An asymmetry appears. Septiles in the natal group (14 E-scientists) show a comparable effect to those in the event group, with the main excess of aspects chiming just *before* exactitude. This is in accords with the traditional astrological notion that applying aspects are stronger than separating aspects:

> An aspect will be stronger when the faster-moving planet of the pair is *applying* to the slower-moving planet, (i.e. coming closer to the exact aspect), than when it is *separating* ...[4]

Though traditional, such a view may seem puzzling, as if an effect could precede its cause. Would one not expect the main effect be *after* the aspect has chimed, as ripples spread out in a pool after a stone is dropped in but not before? A different analogy would be of tension building up as two people walk towards each other, and ebbing away after they have met.

What matters in this data is that which has replicated through the two groups, natal and event. It is far from self-evident that these two different types of group should have much in common, but remarkably this has proved to be the case. Both groups showed a strong septile excess, of peak strength at around one degree of orb, stronger in applying aspects than in separating.

Showing the Septiles

How can septile aspects be recognised in a chart? They don't leap to the eye, and really need a computer chart program to pick them out. It's only fairly recently that programs will show septile aspects. As an example, let's look at the recently-described Edison E-moment, when the electric light concept came to him.

The account we have of Edison's brainwave is as follows:

> Eureka moments and instantaneous insights are part of the lore of invention and discovery ...

EUREKA

On 8 September 1878, Edison experienced a eureka moment when he discussed with the inventor and industrialist, William Wallace, the flawed incandescent lamp system of Wallace's collaborator ... After reflection, Edison found these and related insights of his own so promising that he telegraphed an associate, 'Have struck a bonanza in electric light ...'[5]

It is hard to recall an earlier book on science history which both used the phrase 'eureka moment' and informed readers of the date. This work appeared the year after our 'Eureka Effect' report, so let's hope it is a harbinger of things to come.

Setting the program to display only septile aspects for this date at noon gives the chart shown (figure 7.3).[6] They are formed between Moon and Pluto (not significant, as it's untimed) and Venus, Jupiter and Neptune. Four septiles plus three quintiles were present at this E-moment of Edison, using the Addey orbs.

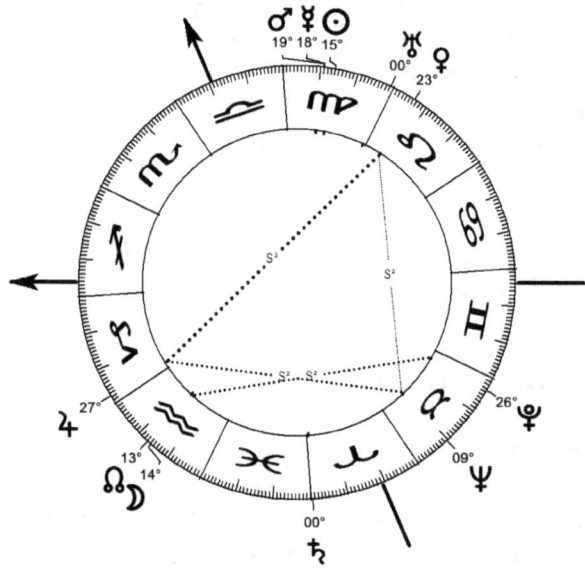

Figure 7.3: Event Chart for Edison's E-moment (septile aspects only) at Wallace's lab on 8 Sep 1878 14:00hrs, New York NY, USA. (Addey orbs used.)

An alternative approach (which we used in the 1980s, in the bad

Chapter 7. Silhouette of the Septile

old days before programs would display septile aspects on a chart), is to call up a 'seventh-harmonic' chart (figure 7.4).[7] This is a technique which moves the planets round, so that all those that are in septile aspects to each other will here appear as adjacent.

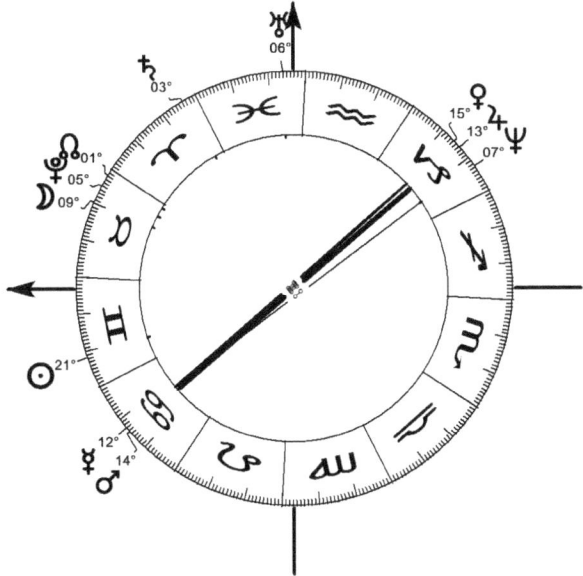

Figure 7.4: 7th-H Harmonic Chart for Edison's E-moment showing conjunctions and oppositions only. (Addey orbs used.)

The septile aspects in this chart appear as conjunctions, while the 14th-harmonic aspects appear as oppositions, and so forth. Only the oppositions are here depicted. There are five or six of these, which makes them at least as strong as the septiles. If this bit about the 'harmonic' chart confuses you, just forget it — it isn't important. It just gives us a means of depicting a higher harmonic, in this case the 14th, as one couldn't otherwise show in a chart. This endorses what was found earlier, that high-frequency 'overtones' of the septiles are significant at these moments.

References

1. Lecture by Brahe at University of Copenhagen, 27.9.1574, *Oratorio de Discipliniis Mathematicis*: Sven Svensson, *Dynamic Astrology*, Spearman, 1983, p.12.

2. Addey: Chapter 1: Introduction, reference 1.

3. Checked using *Frequencies for Aspect Research (FAR) Program*, Mark Pottenger, Int. Soc. for Astrol. Res. (ISAR) LA, California — also the 'Jigsaw' program.

4. Jeff Mayo, *Teach Yourself Astrology*, 1979, p.101.

5. T. Hughes, *American Genesis: A Century of Invention and Technological Enthusiasm,* 1870–1970, Viking 1989, p.75.

6. Solar Fire Gold v7.3.1, Esoteric Technologies Pty Ltd.

7. M. Harding & C. Harvey, in *Working with Astrology*, 1990, go into these 'harmonic charts' in some detail.

Part II
The Moment of Invention

CHAPTER 8

Uranus the Awakener

'O for a Muse of fire that would ascend
The brightest heaven of invention'

Shakespeare, *Henry V*

AFTER THE PLANET URANUS WAS DISCOVERED in 1781, there began the great flow of scientific inventions, which soon developed into a torrent. Astrologers are nowadays remarkably unanimous as regards Uranus being the planet associated with scientific innovation and discovery. A popular astrology book claims that 'Uranus is associated with ... modern science — aeronautics, radio and television, space travel,' and it cites as a Uranian trait, 'inventive'.[1] Is there really such a connection?

The category of invention-moments are those thrilling times when *inventions first worked*. Examples were Edison making his electric bulb light up, Enrico Fermi switching on the world's first nuclear reactor, and William Shockley's team assembling the first semiconductor. This list started out as a bin into which such not-quite E-moments could be placed, but it gradually became evident that they were themselves a group, just as interesting although different in kind. These are a more concrete type of event than the eureka moments: in them, new ideas are *actualised*.

We collected the occasions when inventions first worked, that is to say when some new principle of science or technology first found embodiment. They are not mere moments of first public demonstration, though these are often easier to locate. Should one expect that a group of these moments will show some cosmic pattern? They are generally speaking the most important moments in the inventor's lifetime, points in time and space at which the individual enters the historic process, when the personal act becomes of a universal significance. We seek for that pattern.

Invention and Eureka Moments

As is conception to birth, so are the eureka and invention moments connected. Both types of event refute the words of Ecclesiastes, that there is nothing new under the Sun, being occasions when the future arrives. As such, they are accompanied by euphoria and excitement and contribute to the notion of scientific progress. They are moments of success, as ideas or inventions which did not work are excluded. Oddly enough, they seem only to have occurred in the Northern Hemisphere and to have been made almost exclusively by men. The average age of our inventors at their I-moment was 37 years old, which is much the same as that for the E-moments. The latter do not tend to arrive at a younger age than the I-moments despite strong popular prejudice to the contrary.

They are both genesis-moments, in that new growth began from them, which is why they should be of interest to astrologers. If one takes a large industry such as that of the motor car, and then traces it back in time, one arrives at its nascent phase when there was only one of the invention in existence: in this case it would be Étienne Lenoir driving his car through the streets of Paris in 1862. This seems not to have created a great stir, as no-one recorded its date.

Some inventions enjoyed no single moment of birth. For the steam engine or the motion-picture, the birth process was a gradual series of steps. To quote Robert Stevenson, 'The Locomotive is not the invention of one man, but of a nation of mechanical engineers.' George Stevenson (no relation) probably built the first decent, workable steam engines starting in 1814, which the public were not afraid to get into in case they blew up, but a whole string of predecessors lead up to his work — starting from Cugnot's steam tractor of 1769, which pulled heavy guns, the first self-propelled road vehicle. The Cornishman Richart Trevithic first put a locomotive onto a railway line, but that hardly qualifies as an I-moment.

Chapter 8. Uranus the Awakener

The motion camera developed into cinema through a line of inventors including Prince, Edison and the Lumière brothers, with others fighting for patents. No consensus exists that one of these was its inventor. Here we could take the date of a first public display as our I-moment, which was achieved by the Lumière brothers in Paris, as such times are easier to find, but these are not what we are seeking. To give a more recent example, histories of the Internet give no single moment when it began: rather, at the end of 1970, a computer linkup started to happen.

As was discussed in chapter 2: Quintiles, Septiles and Genius, the two types of moment differ in that the I-moments are not times of insight. Nothing new was realised, except that a process or machine worked. The E-moments have a sometimes transcendent sense of unity with the world-process, as some vital principle is realised, as illumination dawns, linked as we have seen to the septile aspects, the septiles of inspiration. These moments are mould-breaking and shatter established patterns of thought. Though they may have a laboratory or observational context they are in essence 'interior revelations' in that they happen within the mind of the scientist. For this reason they are harder to find, being often cast by the wayside and omitted from the scientist's final report.

The I-moments are public events, witnessed by others if only the laboratory technician, and so are easier to find. Thirty-six such moments were located by the author and M.O.^{2}, as compared with twenty-one E-moments. Their dates are more recent in time: they almost all follow the discovery of the planet Uranus, continuing up to 1987 when the superconductor was developed, whereas the E-moment group started two centuries earlier, with Brahe's supernova of 1572, and our list ended in 1953 with the discovery of DNA. The I-moments have been inspirational experiences for those involved, and as we shall see they have their own septile aspects.

All scientists in the group of E-moments featured in Asimov's *BEST*, with its thousand potted biographies. Thus if someone said, what about including the eureka moment of Klaus von Klitzing, Nobel prize winner, when he found the Quantized Hall effect in Grenoble, at 2 a.m. on a certain date (as Dr Theodor Landscheidt did say to us), then the reply had to be that as the person was not in the Asimov reference, published in 1966, he could not be included. This reference work enabled us to maintain a decent distance from the events. A historian should after all not tread too closely upon the heels of the present (as James I said to Walter Raleigh, as to why the latter was in the Tower).

The *Biographical Dictionary of Scientists/Engineers and Inventors*, here referred to as the *BDS* (not to be confused with the multi-volume *Dictionary of Scientific Biography*) has been our primary source for virtually all inventions of a mechanical nature, though it omits medical inventions such as anaesthesia.³ The Reader's Digest *Inventions that changed the World* of 1982 is a fine coffee-table opus which has also been used as a reference.⁴ A series of invention history books by Egon Larsen helped in determining that often elusive moment when an invention was first seen to work.*

The eureka experiences were located with some confidence, with the help of Koestler's opus, whereas finding the invention-moments was much more of a learning experience. For example, the invention of the neon light by Claude was at first inserted into the list as December 1910, then removed when it was realised that this was a mere public demonstration. The transistor's invention date is normally given as December 24th 1947, the date given in the Reader's Digest sourcebook, whereas it took quite a bit of sleuthing to find the correct date one week earlier, on December 16th, as given in an account by one of the inventors. The first date for radar was initially taken from the *Shell Book of Firsts*, as invented by a German Kuhnold, after which it appeared that invention books generally give the credit to an Englishman, Sir Robert Watson-Watt. A mistaken date for the electric light is given in virtually all source-books, and only quite recently have researchers located what is probably its true date

---

*Despite its title, the Asimov reference could not be used in a like manner for the I-moments, as it omitted too many of the modern inventors — Sikorsky, Whittle, Baird, F. C. Williams (for the first stored-program digital computer). Indeed our I-moment list would have lost the last five and ended at 1947 with Shockley's assembly of the transistor at Bell Telephone laboratories in New York, if we had allowed it to be limited by the Asimov reference. Edward de Bono's *Eureka!* book, for example, was entirely a history of technology and contains no valid E-moments, nor was it very disposed to cite dates. More extensive lists are given in the *Shell Book of Firsts* or the *Guinness Book of Answers*, including such things as the first razor blade, street bollard and so forth, not quite what we are here looking for.

Chapter 8. Uranus the Awakener

and time.

The BDS reference normally gave just one inventor for any given invention, which was convenient. Its use as a reference avoided inconclusive disputes over priority:

- for radio, first working, the credit went to Marconi, not to Mahlon Loomis;

- for the helicopter, to Sikorsky, not to Paul Cornu;

- for radar, to Watson-Watt, not to Kuhnold;

- for the telephone, to Bell, not to Philipp Reis;

- for the jet plane, Sir Frank Whittle, not to Ohain;

- for the hot air ballon, to the Montgolfier brothers, not to Father Bartholomeu Gusmao;

- for the electric telegraph wire, Samuel Morse not to, say, F. W. Cooke, and so forth.

This meant taking the name popularly associated with each invention, as had a very practical advantage: the life of the more celebrated name is better documented, and so the date when that person first made the invention work was easier to locate.

There are rare cases where E- and I-moments are found in the same life, e.g., the idea which came to Nicola Tesla in a flash of inspiration in Budapest Park so that he etched it in the sand as the sun set, in February of 1882; was finally embodied in the throbbing pulse of his 'polyphase' electric motor, on the other side of the world in New York six years later; whence the 50 cycles per second electricity powering our modern world began. One suspects that somewhere in New York, or perhaps confiscated by the FBI, are documents which would enable us to date this epochal moment, when Nicola Tesla set the wheels in motion. The same could be said of Fermi, that his E-moment in Rome, concerning the mode of entry into an atomic nucleus by slow neutrons in 1932, was applied in 1940 in Chicago when he switched on a uranium pile. The life of Fermi will be of especial interest to astrologers, not merely because his birth data is reliably known, but because these two moments fall within that extraordinary decade following the discovery of Pluto when the power of the atom was unleashed.

Apart from these exceptions, however, we are generally speaking concerned with two different types of person, and so have grouped them as *either* an inventor *or* a scientist, the temperaments being somewhat different. Alexander Fleming in London discovered penicillin which Florey and Chain in Oxford, one Jupiter-cycle later, prepared in a usable form. Townes had the idea for the laser which Maiman built, and come to that, Leonardo da Vinci had the idea for the helicopter which Sikorsky built. In that case, there was a gap of four centuries, whereas for the discovery of X-rays, a mere month elapsed before the theory was put into practice.

Inventions are often discovered simultaneously by different persons. Edison invented his phonograph in the same year as the French physicist Charles Cross was exhibiting his talking machine to the Paris Academy. He made his electric light work while Joseph Swan was exhibiting his electric light in Northumberland, and Bell beat a rival for the telephone patent by a mere hour or so. The invention of radar bubbled up simultaneously in half a dozen European countries. These synchronies seem to be quite normal in the history of invention, and yet are rare for the eureka experiences, which enjoy a more solitary splendour.

Mythical eureka experiences blossom, perhaps just from the desire for a good story. Marie Curie gazes in amazement at her sample of uranium glowing in the darkness, separated at last from the tons of pitchblende, with husband at her side ... or at least she did in the film. Her biographer was at pains to stress that no such dramatic moment ever occurred. The Abbé Huay, founder of the science of crystallography, is described in Koestler's opus as suddenly seeing a stone broken open, whereupon the hidden crystal pattern dawned upon him and he rushed home to smash up his own collection of calcite crystals to ponder their inner structure. More sober books on crystallography point out that efforts to trace a source for this story have all been in vain.

Likewise, I-moments are fabricated, though not so often, in part because they lack the glamour, but mainly because of their more public nature. The occasional I-moment will evince a myth-generating capability, for example the story of Cockroft and Walton on the streets of Cambridge in 1932, stopping passers-by to exclaim 'We've split the atom! We've split the atom!'[5] Cockroft and Walton had made their particle accelerator work at the Cavendish laboratory. Scientists tend to be rather Saturnine by temperament and are not given to such displays, at least not in England. Books about Nicola Tesla describe his generation of artificial lightening, at Colorado Springs, climaxing with the fusion of the local power station. However, local records lack any record of such an event.[6]

America — Land of Invention

Within our I-moment group, America scores 15, followed by the UK with 12, and France with 4 (Germany scores zero, but it has a few 'lost' moments: the date for the first electron microscope in Berlin of 1932, for example, presumably incinerated in the War; or the gyrocompass which Aschutz invented in 1907). Thus America has a clear lead in this era of technology. But, if one includes the 'lost' moments (appendix D), many of which tend to be earlier, then, overall, Britain has enjoyed more E- and I-moments than any other nation. National bias could easily creep in here, and one has to be careful. Overall, for moments lost and found, the US has stormed ahead in technological innovation through the last century and a half, say from 1840 to 1990, when it achieved eighteen I-moments as compared to fourteen for the UK.

One day, let us hope that Eureka Tours Inc. will visit the very spot in Rugby tennis club where Denis Gabor was sitting when his hologram concept came to him one Easter morning; or the Eagle tavern in Cambridge where Crick emerged to announce that the DNA structure had been discovered; or the traffic lights in Southampton Row, Holborn, where Szilard had the idea of a chain reaction. Invention moments may not have quite the same glamour, but tours could show the spot in Sheffield where Harry Brierley had the first stainless steel knives made by Mr Ernest Stuart of a local cutlery firm (the date lost); or the Soho attic where Baird first showed a face on a TV set; as well as visiting the Royal Institution at Piccadilly where Faraday made his electrical discoveries. Sadly, many of these historic sites are now concreted over, as their dates have vanished into oblivion.

Table 8.1: Inventions: days when they first worked (n=36).

What	When:Date	Time	Where	Who
Barometer	1648, Sep 19	11:30hrs	France	Perier
Lightning Conductor	1752, May 10	14:00hrs	Paris, France	d'Alibard
Balloon	1783, Jun 4	14:00hrs	Annonay, France	Montgolfier
Vaccination	1796, May 14		Bath, UK	Jenner
Electric Motor	1821, Dec 25		London, UK	Faraday
Transformer	1831, Aug 29		London, UK	Faraday
Solenoid	1831, Oct 17		London, UK	Faraday
Dynamo	1831, Oct 28		London, UK	Faraday
Anaesthetic	1846, Oct 16	11:00hrs	Massachusetts, USA	Morton
Foucault Pendulum	1851, Jan 8	02:00hrs	France	Foucault
Telephone	1876, Mar 10	19:00hrs	Boston, USA	Bell
Phonograph	1877, Dec 6		New Jersey, USA	Edison
Electric Light	1879, Oct 23	01:30hrs	New Jersey, USA	Edison
X-Rays	1896, Jan 17		Vienna, Austria	
Radio Station	1897, May 10	19:00hrs	Cardiff, UK	Marconi
Powered Flight	1903, Dec 17	10:40hrs	N. Carolina, USA	Wright Bros.
Insulin	1922, Jan 11	15:00hrs	Toronto, Canada	Banting
Television	1925, Oct 2		London, UK	Baird
Rocket	1926, Mar 16	14:30hrs	Massachusetts, USA	Goddard
Particle Accelerator	1932, Apr 13		Cambridge, UK	Watson
Radar	1935, Feb 25	17:00hrs	UK	Watson-Watt
Helicopter	1939, Sep 14		Connecticut, USA	Sikorsky
Penicillin	1940, May 25	18:00hrs	Oxford, UK	Florey & Chain
Jet Plane	1941, May 15	19:35hrs	Midlands, UK	Whittle
Nuclear Reactor	1942, Dec 2	15:20hrs	Chicago, USA	Fermi
Atom Bomb	1945, Jul 16	00:00hrs	New Mexico, USA	Oppenheimer
Transistor	1947, Dec 16		New York, USA	Shockley
Computer	1948, Jun 21	00:00hrs	Manchester, UK	Williams
Atomic Clock	1948, Aug 12		Washington, USA	Lyons
Thermonuclear Device	1952, Nov 1	07:15hrs	Elugelab, Pacific	Teller
Sputnik	1957, Oct 4	09:00GMT	Caspian Sea	
Hovercraft	1959, May 30		Cowes, UK	Cockerell
Laser	1960, May 15	10:30hrs	Malibu, USA	Maiman
Hologram	1963, Dec 4		Massachusetts, USA	Leith
Superconductor	1987, Jan 29	17:00hrs	Houston, USA	Chu
Gene Therapy	1990, Sep 14	12:52hrs	Washington DC, USA	Bethesda Hspl.

Local time has been given where known. The principal references used were:
The Biographical Dictionary of Scientists, Engineers and Inventors
and Readers Digest, *The Inventions that Changed the World*.
See the next table for the list as published by *Inventor's World* (1996).

In 1955, at the Teddington National Physics Laboratory, atomic time officially began.[7] Time, hitherto measured by the hourglass, the water closet, the gnomon, the pendulum and the watch-spring, began to be measured by a caesium beam, a beam accurate to one second in a million years, provided no-one turns the electricity off. I inquired regarding when this new era of chronometry began. Did the technocrats of Teddington celebrate the occasion of another great British first? If they did, no-one remembers it. They lost the date, and even mislaid the month when the era of atomic time began. It was in May or June. Such disregard for genesis-moments is, alas, all too common.

Table 8.1 gives all major Invention-moments as could be dated. There's only been one alteration made since we first published this list: the first successful gene therapy treatment, at Bethesda hospital Washington D.C. in 1990 (see chapter 9) has been added in. One doubtful case has been removed, the 'helio-spectroscope' of 1868 (as pointed out to us by physicist S.B. when he kindly checked through the list), this being more by way of a discovery than an invention.

Aspects to Uranus

We decided to test the consensus opinion amongst astrologers about the planet Uranus. Would it be stronger in the usual aspects which they employ? These five aspects — conjunction, opposition, trine, square and sextile — are often referred to as the 'Ptolemaic' aspects, as being the ones described in antiquity by Claudius Ptolemy. For orb we chose five degrees as being the size used by several previous investigators.[9] This represents an average value used by astrologers for orb over these traditional aspects.

One sometimes finds it claimed that Mercury and Uranus together predispose to the inventor's temperament: as Charles Carter affirmed,

> Inventiveness is an outcome of the Mercurial vibration, ... plus Uranus. ... However clever the Mercurial person may be, he does not, without a prominent Uranus, go far beyond revising, arranging, and compiling the work that others have originated.[10]

Table 8.2: Total Uranus-aspects in the group of 36 actual.

	Actual	Expected*	Excess	
Conjunction, opposition and trine:	60	36.7	+61%	$\chi^2 = 13$
All five aspects:	94	73.5	+31%	$\chi^2 = 7$

EUREKA

Figure 8.1: Anniversaries of Invention[8]

But, one cannot really test such a specific concept using a group of merely three dozen dates.

For the untimed moments we took 2 p.m. — after all, one hardly

Chapter 8. Uranus the Awakener

hears of inventions in the morning. On average, the Moon's position would be out by one degree for these untimed moments, if the events were within four hours of this time — which seemed to us as just about acceptable. Table 8.2 shows these Uranus aspects in this group, and compares them with the expected frequencies. The large excess occurs in the first three aspects and most strongly in the trines, but not in squares or sextiles. We see a massive *sixty percent excess of major aspects* to the planet Uranus, in the I-moment group. Table 8.3 shows each aspect separately.

Table 8.3: Strength of the Uranus aspects (for 5° orbs).

Uranus Aspects	Frequencies	Expected
Conjunction	19	9.0
Opposition	12	8.0
Trine	29	19.7
Square	15	17.7
Sextile	19	19.1

Next, taking the first three major aspects at five degrees of orb between all planets in the group, the computer added up the score for each of the planets as shown in the bar-chart (figure 8.2). There was a *general excess of such aspects*. Only Mars and Saturn, the two traditionally malefic planets, scored below chance. The overall expected frequency was found by the computer to be slightly below the theoretical mean of 36.

We begin to discern the 'pattern in the stone', the structure that these moments in time have shared in common. Uranus at the top of the list and Saturn at the bottom does powerfully contrast their influence at these revolutionary moments. How interesting that the two traditionally malefic planets, Mars and Saturn, have both scored below chance at the I-moments. The Sun, which came top of the list of the E-moment aspects, has also done well in the I-moments: perhaps these events are also quite illuminating! But, why on Earth should

---

*The expected frequency for the five Uranus aspects comes out rather conveniently at exactly 2.0 per chart, or just 1.0 if we score only the first three aspects of conjunction, opposition and trine. Very comparable expected frequencies were obtained by the computer from long-period sampling over three centuries (appendix B): 1.02 per chart for the first three aspects, and 2.04 for all five.

EUREKA

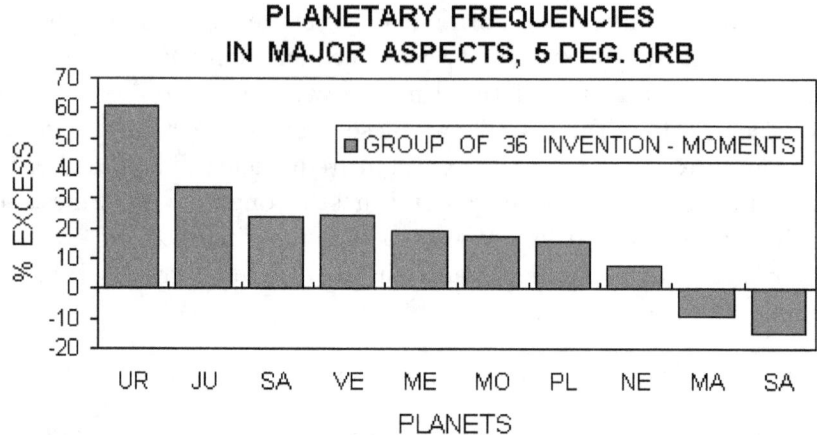

Figure 8.2: Scores of conjunctions, oppositions and trines in the I-moment group.

the excess be in trines and not, say, squares? Later on (chapter 10: Structure of The Trine) we grapple with this question.

References

1. Derek and Julia Parker, *The Compleat Astrologer*, p.101; Dennis Elwell, *The Cosmic Loom* 1987 p.100.

2. N.K. and M.O, 'Invention-Moments and Aspects to Uranus' *Correlation*, 1992; reprinted without essential alteration in *The Eureka Effect*, 1996.

3. Abbot Ed., *The Biographical Dictionary of Scientists: Engineers and Inventors*, 1985.

4. Reader's Digest Association, *The Inventions that Changed the World*, 1982.

5. The story was narrated in Timothy Ferris, *The Coming of Age in the Milky Way* 1988, p.319. For what actually happened, see *Cockroft and the Atom*, Hartcup and Allibone, 1986, Bristol, p.52.

6. See: *Prodigal Genius, The Life of Tesla* by John O'Neill. However, a letter to the author from the Pikes Peak Librarian at Colorado Springs said that they had sought in vain through the 1899 local archives for a mention of the event.

7. P. Robertson, *The Shell Book of Firsts*, 1974.

8. Ref. for Anniversaries of Invention: The Moment of Invention, Inventor's World, Spring 1996, pp.14–15.

9. See, e.g., Michel Gauquelin, 'Astrological Aspects at the Birth of Eminent People' *Correlation* May 1986 pp.25–35 (a negative-result study).

10. Charles Carter, *An Encyclopaedia of Psychological Astrology*, 1924, reprinted 1977, p.114.

CHAPTER 9

When Inventions Worked

THE INVENTION-MOMENTS INVOLVED TRIUMPH and excitement, innovation and achievement. The following accounts locate these dates and whenever possible the times, and point out relevant celestial patterns. They aim to provide enough detail for the reader to judge whether the event is really an I-moment, and, for timed events, whether the optimal time has been chosen.

Historians often describe the person who had a new idea rather than the inventor who made it work, as if the latter were unimportant. We here redress that imbalance, by describing how: Franklin may first have conceived the lightning conductor, but D'Alibard first constructed it; Röntgen discovered X-rays, as were first used in a Viennese hospital a month later; Fleming discovered penicillin, but Florey and Chain in Oxford first produced it in a usable form; Townes had the idea for the laser, but Maiman first built one; Gabor had the idea for the hologram, made to work by Leith and Upatnieks; and so forth.

Thanks are due to the late Ananda Bagley, the director of Electric Ephemeris, for his insightful comments of an astrological nature. He appreciated the limitations of the present study, i.e. of commenting only upon angles and omitting other astrological factors. The orbs he used for these charts are wider than the five degrees we have used for the statistics, and in addition his charts included the minor aspects.

The patterns shown in these charts have aesthetic qualities which help one to appreciate what was happening at these times. In this, they contrast with those of the eureka moments, which tend to have little visual interest for the astrologer. The subtle angles which characterise the latter, the quintiles and septiles, do not show up when one looks at a chart, and it is hard to see much by way of transit linkage with the scientists involved. Possibly because the I-moments are of a more concrete nature, their charts are often of marvellous visual interest, and moreover can be significantly linked to the chart of their inventor. Astrologers who have never mulled over the charts of, say, the lightning conductor, the hot-air balloon, the laser beam, or the computer, have definitely missed something important.

The Uranus aspects have been printed to the right of each title. As explained in the previous chapter, their expected number was two per chart (on a chance basis) for the five Ptolemaic aspects at 5° orb. Readers may prefer to dip into these stories at leisure, rather than read them all at once.

Barometer

1648, 19 SEP　　　　　　　　　　　　　　♅☌♆, ♅☍♀, ♅□☿

Our earliest I-moment takes us to the mountains of central France, where the weight of air was first demonstrated. In 1648, the Italian Torricelli showed that a column of mercury would respond to air pressure, by rising to just 30 inches in a tube sealed at the top, but no higher. From this Torricelli concluded that, 'We live submerged at the bottom of an ocean of air.' A vacuum existed at the top of the tube he said, which was a very heretical view. Air was not supposed to have any weight. Torrichelli sent a letter about this to a Michelangelo Ricci in Rome, who sent it on to Father Mersenne in Paris, who passed on a copy to Blaise Pascal, the mathematician and philosopher. To test this new opinion required a high mountain where a pressure difference would show up. There were none around Rouen where he lived, so Pascal wrote to his brother-in-law Florin Périer in Clermont-Ferrand in the Auvergne, a village surrounded by towering mountains. Would Monsieur Périer kindly perform this experiment?

On September 19th, at sunrise at five in the morning, Monsieur Périer judged that the day was suitable for the great experiment, for he could see the top of the local mountain, the Puy de Dôme, which was normally cloud-capped. He notified various town worthies, clerics

Chapter 9. When Inventions Worked

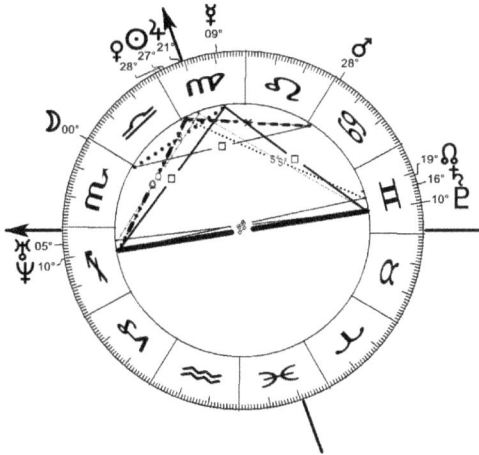

Figure 9.1: Event Chart for Barometer — 19 Sep 1648 NS 11:30hrs local time, Clermont-Ferrand, France. (Addey orbs used.)

and town counsellors, who met together in his garden at eight o'clock in the morning to start the ascent. The account which he sent to his brother-in-law Pascal, our only source for this event, mainly stresses the various ladies who came with them to this morning assembly, which made quite an impression upon him. Monsieur Périer divided his mercury into two lots, having purified it over the previous few days. There were two bowls, each with long glass tubes sealed at one end to allow the mercury to rise up.

The group set out, climbing the 4000-feet high Puy de Dôme mountain with their tube of mercury plus dish. On reaching the top he tested the mercury column at several different spots, under different conditions, and ascertained that indeed the mercury height in the tube was lower than at the bottom of the mountain. 'Everyone was elated. The barometer had been invented,' to quote from James Burke's graphic account in *Connections*.

The group then climbed down the mountain, stopping on the way at a disused church to repeat the experiment. On returning to Clermont-Ferrand, they were assured that the level of the column of mercury in M. Périer's garden had remained the same 'pendant toute la journée.' From this account, we take 11:30 a.m., plus or minus one hour, as the time when they reached the top of the mountain and

145

performed their first trial.

The chart for this event shows Mercury poised to accomplish something quite important. We have taken about 10:20hrs local time, when the group were perhaps still climbing up the mountain, holding the precious mercury. There were three strong Mercury aspects on that day, two of them quite exact, i.e. within less than one degree, plus three Uranus aspects. It only happens once in many centuries that Uranus and Neptune are conjunct, and in opposition to Pluto. Saturn is just coming near to this great opposition, and crossed the horizon around noon, as the experiment would have been in full swing: Uranus, conjunct Neptune, was just rising; the Sun, conjunct Jupiter and Venus, was just culminating; and Saturn, conjunct Pluto, was just setting. It was a mercurial moment with far-reaching implications. The weight of the sky had been demonstrated from a tube of mercury.[1] Périer's birth date has been lost.

Lightning Conductor

1752, 10 MAY

As the Grand Master of the Pennsylvania Masonic lodge, Benjamin Franklin was accorded the highest Masonic honours, and in March 1752, he was appointed to the Committee for building the Freemason's Lodge in Philadelphia. As the leading spirit of the American Philosophical Association, which he had founded the previous decade, he maintained a prolific correspondence with Britain's Royal Society, of an electrical nature. One letter of his sent in 1751 concerned an experiment involving lightning, potentially deadly in its execution. It expressed his controversial new opinion, that lightning *was* electricity. Franklin's essay was translated into French by the physicist J. F. D'Alibard and caused a sensation, after which the latter erected a conductor over his own house. This was a modified version of Franklin's proposed experiment, had some safety modifications, and was a great success.

D'Alibard was not present when his conductor was struck by lightning at around 2:10–2:20 p.m. on May 10th, 1752, in Marly, France. 'Shortly after two o'clock' there came the first rumble of an approaching thunderstorm. Benjamin Franklin acknowledged that D'Alibard was 'the first of Mankind, that had the Courage to attempt drawing

Chapter 9. When Inventions Worked

Lightning from the Clouds to be subjected to your experiments.' A few days later, D'Alibard gave an account of the event to the French Academy. The lightning conductor experiment in France established Franklin's scientific reputation. The opening words of R. W. Clarke's biography of Franklin are,

> On May 10, 1752, there took place on Marly-la-Ville, a small village 25 miles North of Paris, an experiment of crucial importance, not only to science but to the history of the world,

adding, 'D'Alibard's experiment made Franklin famous throughout Europe and forever afterward he was to be lauded by authors and poets as the patron saint of lightning protection.'[2]

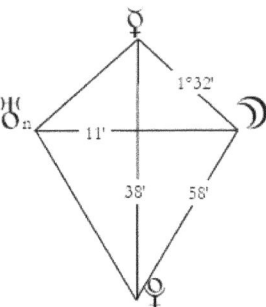

Figure 9.2: Kite formation in the Lighting Conductor chart high in the sky and almost vertical, made by Franklin's natal Uranus (7° 36' Leo) in trine to its Moon (11') and Pluto (41'), and sextile to its Mercury at MC.

Franklin became much better known than the person responsible for the experiment! Some months after this lightning conductor experiment in France, Ben Franklin published a newspaper report in the Pennsylvania Gazette, of which he was the editor, of his fabled kite experiment. No date, place, time or witness was cited, nor was it ever replicated, or even written up in a science journal. It was a self-created myth, rivalling Newton's apple story. As the only scientific experiment ever depicted on US currency, it put Franklin up there with Zeus and Thor in the mythology of lightning.

The implication was, that he had performed it in June of that year, *after* the French experiment but *before* the news of it had reached America on the slow-travelling steamship. Around the time of his report, he invented the lightning rod: 'In September 1752', Benjamin Franklin wrote to a friend, 'I erected an iron rod to draw the lightning down into my house, in order to make some experiments with it ...' Wired up to bells, his lightning conductor pealed away during thunderstorms. Visitors would be alarmed by the way the bells would ring when there was no lightning or thunder, but only a dark cloud passing by. The next year, Franklin's 'Poor Richard's Almanack' explained how houses could be protected by a lightning conductor.

'Franklin invented the lightning-rod, the hoax and the Republic' to quote Honoré Balzac (*Bolt of Fate*, p.70). Franklin was master of the hoax as practical joke, and his illegitimate son, William, tended to be involved in them: paintings of the kite flying in a storm feature the 'ever-silent William' by his side. The crux of this impossible experiment concerned a non-conduction by silk thread of the substance of lightning.

From the kite up in a stormy sky, a hemp thread led down to a key, and from this key, a short length of silk thread then passed under a door or into a window, to a hand which thus remained dry. Don't ask how one flew a kite from inside a window or door, nor how the heavy key was stopped from touching the ground. The lightning's electricity made the key sparkle so that 'electric charge' could be drawn from it. In the months after Franklin's report, a Russian professor was killed in St Petersburg, by carefully replicating the experiment.

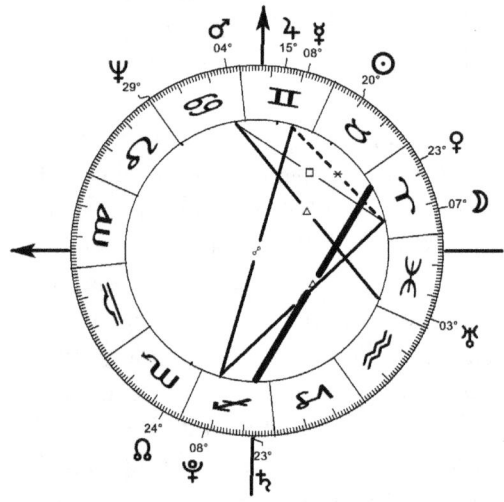

Figure 9.3: Event Chart for Lightning Conductor, 10 May 1752 NS 14:15hrs local time Marly la Ville, France. (5° orbs used.)

Around this time, it happened that a Father Divis of Moravia was

Chapter 9. When Inventions Worked

also setting up a lightning conductor, near his vicarage in the town of Primetice, at an unknown date. Divis wrote to the Austrian emperor recommending he try the new device, but the Emperor was advised against it.³ A drought then struck Moravia and locals believed it was caused by the diabolical and unnatural contrivance which Father Divis had erected. So they tore it down and begged Father Divis not to erect it again. He consulted with his superiors, who advised him not to alarm his flock. What might have been the world's first lightning conductor was thereby laid to rest.

We should expect to find some connection between the chart of Franklin and that of the lightning conductor. Commonsense might decree otherwise, because the two charts were from opposite sides of the world, and Franklin had no say in the date when the lightning conductor experiment was performed. Let us look and see.

When the world's first lightning conductor was struck by lightning, Pluto was forming a close trine to the Moon (29′) and opposing Mercury (34′), so that the Moon and Mercury had a sextile link (1° 4′). That all seems symbolically appropriate enough. In addition, both Pluto and the Moon were in trine to the Uranus of Benjamin Franklin (41′ and 11′ respectively), and so a precise grand trine was then forming, just over the moment of the lightning-fork. Franklin could not have chosen a more appropriate moment. The main axis of this chart, its Pluto-Mercury opposition, was almost vertical as the lightning struck, i.e. it was nearly perpendicular to the horizon. Jupiter, wielder of thunderbolts, was not far from the MC.⁴

Balloon

1783, 04 JUN ♉︎♂♀, ♉︎☌♄, ♉︎△♂, ♉︎□♆

The two Montgolfier brothers, Étienne and Joseph, grew up near Lyons. Joseph Mongolfier worked in the family paper factory while dreaming interminably of flying through the clouds. One day in January of 1772, a notion occurred to him, possibly while sitting in a mail coach travelling to Avignon to visit a customer.

Years later, the townsfolk of Annonay nearby started to hear of a strange device that the brothers Joseph and Étienne Montgolfier were constructing, and eventually the date for a public demonstration was fixed, as June 4ᵗʰ, 1783. 'How often Joseph and Étienne tried the 4ᵗʰ

149

June model before the public demonstration is uncertain,' Gillespie commented, 'but there were at least two or three preparatory launchings.' The first was 'probably the 3rd April' and the second was on the 25th. Despite this, we decided to take June 4th as the first balloon flight, because Gillespie refers to this flight as 'this progenitor of all balloons'.* This sounded like a genesis-moment.

Figure 9.4: Hot Air Balloon

It was drizzling with rain on the day, but the demonstration had to go ahead because of all the local dignitaries who had been invited to come and see this event. Stands were built for them in the market place of Annonay. In the middle of the square was a wooden scaffolding from which an empty linen sack, lined with several layers of paper, hung limply. There were official delegates plus thousands of people crowded round the barriers, filling the windows, and sitting on the house-tops. No-one seriously imagined that the thing would really sail in the air. The two brothers directed four husky labourers, who affixed masts and pulleys, and prepared the brazier underneath.

The four workmen from the factory caught hold of the ropes that hung from the envelope. In the centre of the scaffolding on the ground, suddenly the sack began to fill — it was impossible to see how or with what. The four men struggled to hold on to the balloon, then Étienne gave the command 'Let go!' and the balloon rose quickly. It rose until it was no more than a speck in the sky — almost 6,000 feet high. The spectators, with wild excitement, streamed out of the town after the balloon as it went before the wind. They found

---

*The balloon date is often cited as June 5th, 1783. However, science historian C. Gillespie's *The Montgolfier Brothers and the Invention of Aviation* (1983) established the date as Wednesday, June 4th.

Chapter 9. When Inventions Worked

it a mile and a quarter away — a limp, empty sack as before.

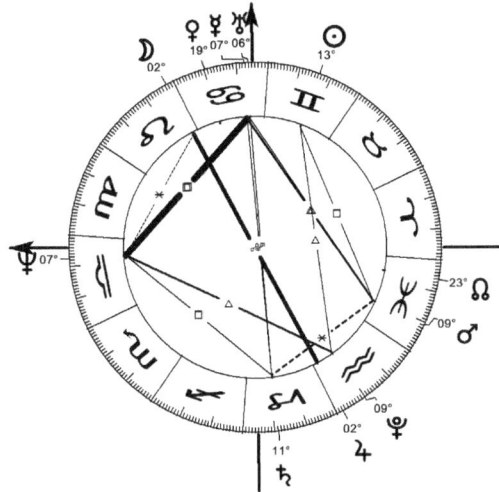

Figure 9.5: Event Chart for First Balloon Flight, 04 Jun 1783 NS 13:30hrs local time Annonay, France. (5° orbs used.)

Gillespie tells us that 'people began gathering late in the morning' waiting for the event. From this we may conjecture a launch time between 1 and 2 p.m.

A report was sent to Paris of the sensation, referring to the mysterious 'Montgolfier air' which had lifted the balloon up. Some years later, the Frenchman Charles was to invent his law concerning the expansion of gases when heated, but for now it was a complete mystery. The Paris Academy considered the matter. Louis XVI wished to see a balloon ascend, and fixed September 19 for the event. The Montgolfier family all agreed that Joseph was too shy and unworldly to go to Paris, so his brother Etienne was sent.

They constructed a new airship of cotton and paper, with a cage suspended to hold a duck, a cock and a sheep. A cannon was fired at one o'clock on the appointed day, then a second one at 11 minutes past one when it ascended. With its bleating, quacking, crowing passengers, the balloon stayed aloft for eight minutes. Then a hole grew in its side, and it became entangled in a tree.

Later that same year the first human flight took place. The first 'free' human flight with no ropes constraining it took place on November 21, 1783, shortly before 2 p.m., from the Bois de Boulogne. A craze for ballooning swept through France.

Joseph Montgolfier had Neptune conjunct Jupiter (2°) opposition Uranus in his natal chart. The rather dreamy and imaginative planet Neptune symbolises the aspiration which the young Joseph had to leave the earth and drift amongst the clouds. On the day of his great demonstration in his home village, Neptune rising stood in four different aspects. A strong Mercury-Uranus conjunction (2°) stood at the MC in opposition to Saturn at the IC. That conjunction aligned with Montgolfier's Jupiter-Neptune conjunction, and the Saturn in opposition was precisely conjunct (27′) his natal Uranus! This event happened during his Uranus-opposition-Uranus return.

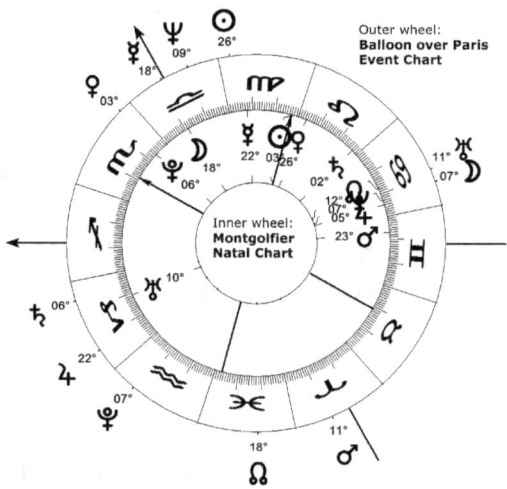

Figure 9.6: Transit Chart for Balloon over Paris [outer wheel], 19 Sep 1783 NS 13:11hrs local time Paris, France; to Natal Chart for Joseph Montgolfier, 26 Aug 1740 NS Annonay, France [inner wheel]. (5° orbs used.)

Moving on to that historic day when he demonstrated his skill to the world, when before the astonished gaze of Louis XVI and the people of Paris some farm animals rose into the sky, we see even stronger Neptune aspects, five of them, including squares to Uranus and Saturn. It was the day of his Uranus half-return to within 2′, which means that the event was within a day or so of his exact Uranus half-return. The Moon and Uranus were conjunct, still over Montgolfier's natal Jupiter-Neptune conjunction, and Saturn in opposition now a few degrees from the natal Uranus.

Chapter 9. When Inventions Worked

Figure 9.6 shows Montgolfier's Uranus opposition return plus Neptune square return synchronising, together with a Pluto square return at his balloon launch; also, the Uranus-Saturn opposition then present aligning with the Neptune-Uranus opposition in his natal chart.

Both balloon charts had grand trines, in the air signs (see figure 9.7). (A 'grand trine' is an equilateral triangle, whose three positions will normally fall in zodiac signs of the same element. The three 30° divisions of the zodiac air signs are spaced 120° apart.) Neptune was trine Pluto in the first balloon chart with the Sun forming the third point of the triangle. In the second rather beautiful balloon chart, when the three animals ascended with all Paris watching, it was the Moon which formed the third point of the triangle.

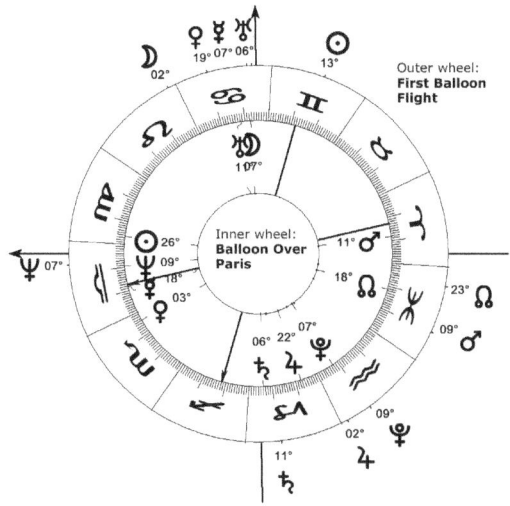

Figure 9.7: Event Chart biwheel for first balloon flight [outer wheel], 04 Jun 1783 NS 13:30hrs local time Annonay, France; with balloon over Paris [inner wheel], 19 Sep 1783 NS 13:11hrs local time Paris, France. (5° orbs used.).

The unusual structure to the two balloon charts, which occurs only once in centuries, is the opposition of Saturn and Uranus, with a T-square to Neptune, i.e. Neptune in a 90° angle to this opposition in the second chart. The two charts are notably well-structured: in the first (outer wheel), Mercury conjuncts Uranus to reinforce the T-square, with Mars in trine to Mercury-Uranus and in sextile to Saturn. In the second (inner wheel), Mercury is again exactly

153

conjunct the MC. This chart is exactly timed whereas the first is not, it is more meaningful in this case. Mercury has moved 90° between the two charts: in the first it was conjunct Uranus and in the second, Neptune. Mars has moved on 30°, so as to oppose Neptune (Mars opposition Neptune 3°, square Uranus 1°) forming a perfect grand cross. It squares Uranus (1°), whereas it was in trine before (3°). Mercury and Mars have moved in a harmonic manner between the two balloon launches, to reinforce the T-square. As the hot air balloon was powered by fire, Mars is symbolically appropriate, and Mercury at the top of the charts needs no comment.

The invention concerned a gravity-defying upward motion. The two charts both focus on the MC. (the 'Medium Coeli', or middle of the sky). In the second chart, the grand cross and the grand trine are linked together by Mercury, which is right on the MC, and the first chart has an equally strong focus at that point. Similar remarks apply to the first powered flight by the Wright Brothers, which we come to later on.

****

When reports of the Montgolfier's stupendous achievement in June reached Paris, one popular French scientist offered to build a balloon himself, as the Parisiens were very impatient. Monsieur Charles leapt to the mistaken conclusion that the secret of the special air which the Montgolfier brothers were using was hydrogen, and at grave risk he began filling a huge balloon with this potentially explosive gas. This took several days. When ready, the balloon was carefully transported to the Champ de Mars, where the Eiffel Tower now stands. On the afternoon of 27th August a cannon shot rang out at five o'clock as Professor Charles gave the command to release the balloon. It shot

Figure 9.8: Montgolfier Portrait.

Chapter 9. When Inventions Worked

up with great speed, cheered by 300,000 people, half of Paris. 'An indescribable feeling of amazement and enthusiasm filled the onlookers,' said a report. Up and up it floated, out of sight, and eventually exploded in the upper atmosphere.

Secretly, Étienne Montgolfier had been watching, and now advised the French Academy that hydrogen balloons were too dangerous, and only hot air balloons should be allowed. A fierce competition broke out over the two rival modes of flight. Later in that decisive year, Charles built a greatly improved hydrogen balloon, and on the first of December he was all set to go.

The balloon had a net over its top to stop it exploding, an elegant gondola to carry two people, and some ballast for controlling height. A huge crowd assembled, and had bought tickets at exorbitant prices. According to Larsen, a police officer then appeared and said that the king had prohibited the ascent because hydrogen was too dangerous. However, Charles was not one to be deprived of his historic glory, especially with his Mercury transit on that very day (Mercury conjunct Mercury 1°), and he sent back a message that he would shoot himself and take his secret to the grave if he were not permitted to ascend. An hour went by, then the messenger returned with permission, and the ropes of the 'Charliere' were cast off. That tale may be untrue, but if so it shows how readily myths grow around these epic events.

Earlier balloon launches in Paris had been marred by the vast crowds threatening to become unruly, forcing a premature launch, so this time a small pilot balloon was sent up to indicate wind direction and generate an air of expectancy. Monsieur Charles felt, he reported, not the slightest anxiety at lift-off. To quote Gillespie, 'a kind of universal amazement and silence fell upon the crowd. Nothing,' wrote Charles a few days later, 'would ever equal that "moment of hilarity" he felt on leaving earth.'

This voyage was a milestone in the history of aviation. M. Charles and a passenger drifted at medium height for two hours, then he descended and made a perfect landing — to let his passenger out. Fascinated by the thrill of flying, Charles soared up above the clouds to ten thousand feet, until his ears started to ache, and then came down. He described what he felt up above the clouds:

> I stood up in the middle of the gondola, and lost myself

in the spectacle offered by the immensity of the horizon. When I took off from the fields, the Sun had set for the inhabitants of the valleys. Soon it rose for me alone, and again appeared to gild the balloon and gondola with its rays. I was the only illuminated body within the whole horizon, and I saw all the rest of nature plunged in shadow. Amid inexpressible delights, this ecstasy of contemplation ...

If there is doubt about using the Montgolfier brothers' June 4th demonstration as the I-moment for ballooning, one can take the hydrogen balloon moment of August 27th, as this was quite unequivocally the first gas balloon flight, whereas practice balloons were sent up by the Montgolfier brothers prior to their public demonstration. In both cases, four Uranus aspects were present so it would make no difference to the score.

That decisive flight eclipsed the hot air balloon — the 'Charliere' had won the day. Everyone talked of ballooning. Goethe witnessed an ascent at Cassel, and wrote enviously: 'Air balloons have been invented. How near I was to this discovery! Somewhat chagrined at not having made it myself.'

When the first gas balloon ascended on December 1st, there were four major Uranus aspects in the sky (Square Neptune, Square Mars, sextile Mercury, opposition Saturn), and surprisingly there were four rather similar Uranus aspects present on Charles' second flight (Square Neptune, Square Mars, trine Moon, opposition Saturn). Mars was a mere seven degrees away from its earlier position, due to its retrograde motion, so it was in square to Uranus on both occasions. Charles had his Mercury return on that very day (Mercury conjunct Mercury 1°), and he was also having a 'node return', being 37 years old. There was an exact Uranus opposition to Saturn on that day (7′), so it is not surprising that the voyage was a milestone in the history of aviation. As regards Charles' oceanic feeling during the flight, we note the Sun-Neptune sextile (2°).

The Monk of Lisbon

There was a much earlier manned flight performed by a Spanish monk, Father Bartholemeu Lourenco de Guasmão. The reward for

Chapter 9. When Inventions Worked

his one and only flight was to be hunted down by the Inquisition on a charge of sorcery. The records were destroyed and memory of the event buried in a deep oblivion. Until fairly recently, the story was viewed 'as fabulous', so wrapped around with folklore had it become. The story shows that in some sense an invention cannot arrive before its due time.

Born in 1685 at Santos, in the Portuguese colony of Brazil, Guasmão entered the Jesuit order and then obtained a degree as doctor of physics and mathematics at the university of Coimbra. He settled in Lisbon, where the royal court resided, and sent a petition to the King of Spain about his invention. On April 19th, 1709, Doctor Guasmão was granted a patent for his invention plus a professorship at the University of Coimbra and an allowance for life. An act was passed declaring that anyone who used the apparatus without the inventor's permission would incur the death penalty — which may be still in force as it has not been repealed!

In 1709 on August 8th, the guests assembled in the royal castle at Lisbon to witness the maiden voyage of a flying ship. Accounts of the device have a Baron-Munchhausen flavour to them, with bellows, a sail, and stones of amber to absorb the Sun's heat. It seems that Guasmão's flying machine had a dozen or so balloons supplied with hot air through tubes. It lifted off and flew out of sight, with all Lisbon watching, and landed. A rumour was put about that Guasmão was mad and that great evil would come into the world through his diabolical machine, and even the king was turned against him. The inventor destroyed his airship and burned the plans. In the end he fled from Lisbon after being arrested on a charge of sorcery. He died in poverty and for two centuries his name was forgotten, the first man to fly.

The chart for this first manned flight is unusual and worth looking at even though it is untimed. It has Uranus conjunct Pluto and Venus, in trine to Neptune and sextile to Jupiter, so that Jupiter and Neptune are in opposition. Moon conjunct Mars stands at the midpoint of the sextile, i.e. they are in semisextile (30°) aspect to Uranus and Jupiter. Thus the chart resembles the previous balloon charts in that almost everything is in aspect. Saturn is not in any special aspect, which seems appropriate as the event was of no scientific significance.

We use the date of 1783, because Montgolfier is cited by the Biographical Dictionary of Scientists for its invention and not the Spanish monk Guasmão. That date was the historical seed-beginning from which the phenomenon developed, the earlier one having been a premature blossom, remembered as fable. Both dates are well-endowed with Uranus aspects. The earlier one may have the hour of lift-off recorded somewhere in a Lisbon archive, in which case it would become of great interest.[5]

Vaccination

1796, 14 MAY   ♅☌☽, ♅☌♃,[†]♅△♂, ♅□♀, ♅✶♆

In the eighteenth century children normally succumbed to smallpox before the age of ten, and the mortality rate of its epidemics could well be 30 percent. It was indeed 'The most terrible of all the ministers of death' (Macaulay). Or, it left awful disfigurement which some feared worse than death. Edward Jenner, taking his medical apprenticeship at Chipping Sodbury in 1770, heard a dairymaid remark concerning smallpox: 'I cannot take that disease, for I have had cowpox.' Dairymaids sometimes contracted the disease cowpox from the udders of cows, after which they did not contract the disease smallpox. Jenner practised in his native village of Berkeley in Gloucestershire. In a small village one could keep a track of just who had had cowpox and who caught smallpox.

After pondering the matter for many years, Dr Jenner finally decided on May 14th, 1796 to inoculate the arm of a healthy eight-year old boy with pus taken from a cowpox sore on a milkmaid's arm. The boy developed a similar sore but did not become ill. Then, six weeks later, the boy was inoculated with pus from a smallpox sore. Had the boy died or become badly ill, Jenner would have been branded a criminal, but the smallpox did not touch the child. The mild cowpox infection had somehow protected him against smallpox. In 1798, Jenner published his results; in 1803, the first vaccination clinics were opened; and by the mid-nineteenth century smallpox had virtually been eliminated from Western Europe. In 1980, the World Health Organisation declared smallpox to be extinct.

[†] Just outside 5°orb

Chapter 9. When Inventions Worked

Vaccination may first have been applied in 1771 by the surgeon of Bridport Mr Downe, to induce cowpox in a local butcher as a protection against smallpox, but there is no date for this. This and other priority claims did not come to light until well after Jenner had published his monograph.

The modern smallpox vaccine is not found in cows, and experts are puzzled as to just what Jenner's 'vaccine' amounted to. Prior to his work, 'inoculation' was a widely used technique which meant the deliberate inoculation of a trace of smallpox virus into the skin, for the purpose of preventing the disease. A hundred years before the existence of a virus was demonstrated, the word just meant a poison which could be transmitted. Bovine cowpox was rare and Jenner had difficulty getting supplies. One of the greatest ever medical controversies raged for a century over this issue, and now that smallpox has vanished it may never be resolved.

Two years after his first risky experiment, Jenner tried again. He obtained some material from one who had supposedly contracted 'cowpox' from a horse, and injected it into a five-year-old child, John Baker. That child died, a fact which Jenner initially omitted to report and for which he was subsequently much criticised. Jenner's surmised, 'May it not, then, be reasonably conjectured, that the source of the smallpox is morbid matter of a peculiar kind, generated by a disease in the horse ... ' The alternative view was that his injection was merely the smallpox virus in an attenuated form, i.e. that his technique was not substantially different from the 'inoculation' practised before him. The controversy raged on, even as the disease was being conquered.

The staggering five Uranus aspects present on the day of Jenner's first trial strongly argue that Jenner's technique *was* the valid new initiative he claimed. The I-moment was within three days of his birthday, so the transiting Sun was conjunct that powerful natal Sun, but also conjunct the close Mercury-Venus conjunction in the natal chart. Mercury was just as strongly aspected in his natal chart, being in six aspects. Transiting Mercury, seven degrees away from the Sun at the time, i.e. much the same distance as the natal Sun-Mercury, was also conjunct the natal Sun. Thus there were both Sun-Mercury and Mercury-Sun transits, which seems symbolically appropriate for the communication of something life-giving via the blood.

Electric Motor

1821, 25 DEC   ♅☌☉, ♅☌♆, ♅△♂

One can only wonder what a deeply religious man like Michael Faraday was doing in his laboratory on Christmas day of 1821. That was the day when the first primitive electric motor began to revolve. Voltaic batteries supplied the current to a copper wire apparatus afloat in a dish of mercury. To quote from his biographer:

> When, after months of work and many ingenious contrivances, the wire began to move round the magnet, and the magnet round the wire; he himself danced about the revolving metals, his face beaming with joy as he exclaimed, 'There they go! There they go! We have succeeded at last!'

He then proposed to his attendant a treat, and they went together to visit a theatre. This was the first of Faraday's mighty achievements in electricity. European scientists had been thrown into a ferment by Ørsted's announcement that an electric current produced a deflection of a compass needle, and this was Michael Faraday's response to the matter. (Ørsted's eureka moment when he noticed that a current deflected a compass needle, made in the winter of 1819/20, is irrevocably lost).

Born in September 1791, son of a blacksmith, the 30-year-old Michael Faraday was just having his Saturn return at the time, when Saturn returned to its position at his birth. It was after this invention that Sir Humphrey Davy began to grow jealous of his young employee. The dire fact that his young laboratory technician was destined to eclipse his own reputation dawned upon him, and he made unseemly accusations of plagiary.

The metals used in this experiment were copper and mercury, so it is of interest that a close Venus-Mercury sextile (3°) had formed. Mercury was also in trine (3°) with Saturn and Jupiter which were then conjunct (2°). The Sun, Uranus and Neptune are all conjunct within 1°.

The first practical electric motor, demonstrating the principles on which subsequent motors have been based, was made in 1835 by Thomas Davenport, an American blacksmith from Rutland, Vermont, who used it to drill holes through steel. It was not a commercial success, being too expensive.

Transformer

1831, 29 AUG   ♄☌♃, ♄△♀, ♄⚹♀

The year 1831 was the turning-point in Faraday's career, when he discovered the technique of electromagnetic induction, on which the whole edifice of electrical engineering rests; that is to say, the link between electricity and magnetism. In August, he wrapped a copper wire around one half of a circular iron ring, and connected it to a battery, and wound another wire around the other half, connected to a galvanometer. It didn't work as he had expected, for there was no continuous flow of current in the second, but only a momentary flow as the current was switched on or off in the first circuit. That was a puzzle. Using this first transformer, Faraday began to visualise his lines of force, which electricians have used ever since.

In front of the Institute of Engineering in London, overlooking the Thames, stands a bust of Michael Faraday, and on the plinth is inscribed, 'Faraday discovered the principle of electromagnetic induction on August 29, 1831.' This is our sole instance of an I- and E-moment being identical.

Faraday was just coming up to his Uranus-opposition return, and the experiment was performed over an exact Sun-Saturn conjunction.

Solenoid

1831, 17 OCT   ♄☌♃, ♄△♀, ♄△♀, ♄⚹♀

For years Faraday had hoped to generate electricity by mechanical means, but he had been mainly occupied by chemical experiments. He finally succeeded on October 17th, using a cylindrical coil of copper into which a bar magnet could be inserted. Each time he thrust this magnet into the coil a current was induced making the galvanometer needle spin round, then when he withdrew it, a current registered in the opposite direction. The original is on display at the

Royal Institution, or Michael Faraday House, off Piccadilly, where he worked.

A Mars-Venus conjunction stood in the sky on that day, recalling the root meaning of the word 'conjunction.' We have no time of day information for any of these experiments, but at 5 p.m. there was a Moon opposition Saturn which would have been quite a good time for him to be doing this. The Mars-Venus energies were utilised to create an alternating current, as the iron magnet moved in and out.

Mars on that day was within 7° of Venus, and on the other side of Mars stood the Sun, also 7° away, which is surely a fine image for the moment when a new source of electrical energy appeared. Mercury was conjunct Venus, and Pluto is in close opposition (2′). Mars and Venus were both in trine to a Jupiter-Uranus conjunction. Transits to Faraday's natal chart reinforce the Mars-Venus polarity: the Venus was conjunct his natal Jupiter to within 18′, and the Mars was opposite his Saturn to 51′.

Dynamo

1831, 28 OCT ♅☌♃, ♅☍☽, ♅△♀, ♅✶♀

A continuous electrical current was finally generated later that month in what Asimov described as 'probably the greatest single electrical discovery in history.' (*BEST*, p.229) Faraday rotated a copper wheel between the poles of a magnet and a current was thereby produced. Venus was still in trine to the Jupiter-Uranus conjunction and Mars was conjunct Mercury.

Faraday wrote up this hectic series of experiments in his notebook under the title, 'Experiments on the production of electricity from magnetism.' Eddy currents in the disc made the first dynamo very inefficient, but soon improved models were available. It was a long time before the dynamo became a practical proposition, but by 1875 a dynamo was installed at the Gare du Nord in Paris supplying electricity for arc lamps.

Chapter 9. When Inventions Worked

Anaesthetic in Surgery
1846, 16 OCT ☿⚷♀, ☿⚷♂, ☿⚹♃

The story of anaesthesia begins in 1799, when Sir Humphrey Davy lectured on 'laughing gas' or nitrous oxide at the Royal Institution. Its use became fashionable at parties, where it promoted conviviality. In America in 1842, a chemistry student William Clark in Rochester gave ether to a woman who had a tooth extracted painlessly under its influence. In September 1846, the Boston dentist William Morton also removed a tooth using ether.

> The patient declared that he had felt no pain during the operation and was discharged well December 7th. Knowledge of this discovery spread from this room throughout the civilised world and a new era for surgery began.

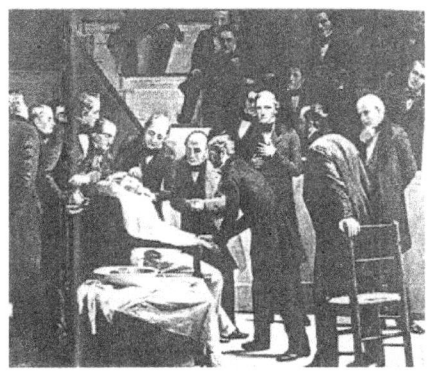

Figure 9.9: Anaesthesia Used in Surgery at Boston

A plaque in a surgery room in Massachusetts General Hospital reads:

The plaque commemorates the epoch-making surgical operation performed under ether on October 16th, from which the practice of anaesthesia developed. The surgeon asked William Morton to use 'the preparation which you have invented to diminish the susceptibility to pain.' Morton administered the ether, and the success of this operation, to quote Asimov, 'once and for all divorced the surgeon from the torture chamber.'

In his dental practice Morton had tried various pain-killers, such as brandy and laudanum, to diminish the agony of the dentist's chair. In the summer of 1844, a Dr Charles Jackson suggested to him that he try sulphuric ether. Morton achieved varying degrees of success with it over the next eighteen months, a substance declared by knowledgeable experts to be a deadly poison. Morton's dental partner, Wells, favoured nitrous oxide, giving a public demonstration of its efficacy in tooth removal, but this was a notable failure.

By the summer of 1846, Morton was achieving only partial success, and surmised that this was due to the ether not being pure enough. Again, he went to Dr Jackson, to whom he had submitted himself as a student of medicine, for advice — a move he was later to regret, when Jackson stepped forward to snatch the laurels of victory. Jackson described to him how the ether could be further purified, after which Morton became convinced he was on the trail of success.

On September 30, Morton experimented with ether, sending himself into unconsciousness. This was a mere week after Neptune had been discovered, on the other side of the world! Morton at 27 years of age was then passing through his Neptune-sextile, within 16' on this date. 'As I recovered I felt a numbness in my limbs with a sensation like a nightmare and would have given the world for some one to come and rouse me. I thought for a moment I should die', his diary records, as he returned to consciousness. At 9 p.m. on that day, when Saturn was conjunct Neptune (1°), a patient appeared who expressed his readiness to try the pure new ether while having a tooth removed. That was the first day when a sound method of anaesthesia was applied, in that the ether was of sufficient purity.

Morton was due at the Massachusetts hospital at 10 a.m., but he arrived late: 'some time was lost in waiting for Dr Morton, and ultimately it was thought he would not appear,' said the hospital report. He was out purchasing some equipment, quite unnecessarily as it turned out. Let us surmise that the operation began around 11 a.m.

As the hospital continued to use the new procedure, the awesome magnitude of the achievement of removing pain from surgery dawned upon the world, and both Dr Jackson, a quite renowned figure at the time, and Mr Wells, Morton's co-partner in dentistry, stepped forth to take the credit. Dozens of pamphlets were published and it was years before the confusion died down, by which time William Morton was ruined and heavily in debt. Astrologers will not be short of comment on the Neptunian significance of these events.

In fact, Jackson had given nothing but advice, and Wells had advocated nitrous oxide, which plainly did not work. The Paris Academy of Sciences, for example, at first fully accepted Jackson's version of events, but then was prevailed upon to award equal honours

Chapter 9. When Inventions Worked

to both claimants, which Morton refused to accept. The US House of Representatives appointed a committee to investigate the issue. Every letter and every testimony was scrupulously examined over a period of eight years. Realising that Morton was getting nothing from his discovery, the US Congress proposed to award him $100,000. However, the lawyers of Jackson and Wells blocked this move and he never received it. Posthumously, Morton received due credit as the inventor.

The operation at Massachusetts General Hospital was *not* in fact the first time a surgical operation had been performed under ether. Four years earlier, a Dr Crawford Long removed a cyst from the neck of a student on March 30th, 1842, in the town of Jefferson. He performed a number of such operations under ether, but desisted when a deputation of local citizens threatened to lynch him. They suspected him of sorcery.

His pioneering role remained entirely unknown until years after the practice had become generally adopted. (This did not prevent a statue from later being erected in his honour in Washington's Hall of Fame).

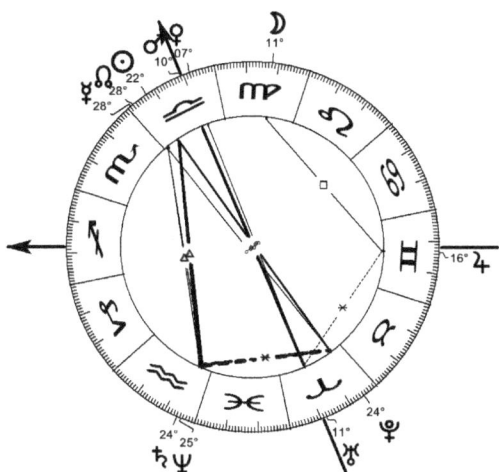

Figure 9.10: Event Chart for Anaesthesia, 16 Oct 1846 NS 11:00 LMT, Boston, Massachusetts, America. (5° orbs used.)

The news of the operation performed in Massachusetts General

Hospital spread rapidly. In December of that year, Morton's method was tried on a tooth-removal operation on a Miss Lonsdale in a London hospital. Britain's first dental extraction under anaesthesia lasted three minutes. The patient reported afterwards that she had experienced no pain, only 'a heavenly dream.' Also in December, one of Britain's foremost surgeons, Robert Liston, at University College Hospital, amputated a patient's leg using ether and afterwards exclaimed memorably, 'This Yankee dodge beats Mesmerism hollow!'

An account by Professor Roberts in his book, *Serendipity*, found that there was 'perhaps no other' discovery of science 'in which credit and honour were so difficult to assign' (p.40). This deep uncertainty reflects the vague and numinous quality of the new planet, Neptune.

Astrologers have always associated Neptune with the practice of anaesthesia. The new planet swam into human ken on September 23, 1846, three weeks before the historic surgical operation at Massachusetts hospital under ether. Astrologically, the reason why Dr Crawford Long's practice of surgery under ether did not catch on is very simple: Neptune had not been discovered! The date for the operation at Massachusetts Hospital saw Saturn conjunct the new planet Neptune in the sky (1°), as well as Sun conjunct Mercury, both being in trine to Neptune/Saturn — echoing the trine in Morton's own chart.

The anaesthesia chart is shown (figure 9.10). Neptune is the most heavily aspected planet with *seven* aspects — though, these charts have been set with wider orbs and more minor aspects, than in our scoring procedure. Using 5° orbs, then Neptune would score four of the Ptolemaic aspects, which is still quite a lot: conjunct Saturn 1°, trine Mercury 3°, trine Sun 3°, and sextile Pluto $\frac{1}{2}$°. Dramatically, Uranus is conjunct the IC, which seems appropriate for an invention involving unconsciousness, and Jupiter is just setting.

Morton's natal chart has even more Neptune aspects, ten using the wide orbs as for the chart-wheel diagrams, or four using our 5° orbs: square to Saturn conjunct Pluto, and trine the MC conjunct Sun. We also notice his Neptune conjunct Uranus to $5\frac{1}{2}$°. These are certainly appropriate for the man who established anaesthesia, who deliberately sent himself unconscious using what was regarded as a deadly poison, to establish the principle of surgery without pain. Alas, they are also appropriate for the miasma into which his life

Chapter 9. When Inventions Worked

entered, the terrible legal quagmires, the confused claim and counter-claim over priority which dogged his life.

Morton's natal Sun at 25° 37′ ($\frac{1}{2}$°) Leo is in close opposition to the focal point of the anaesthesia chart, its Saturn-Neptune conjunction, where Neptune is at 25° 30′ Aquarius. His Neptune forms an exact (15′) semisextile (30°) with Jackson's natal Neptune, so that Jackson's Neptune was in a square to that of the anaesthesia chart. The stars are telling us that both of these individuals had claims to the discovery, in that both had strong Neptune synastry to that chart.

Foucault Pendulum

1851, 08 JAN ♅☌♀, ♅☌♃

Jean Bernard Foucault the French physicist tried to photograph the heavens, fixing his camera so that it rotated slowly with their motion. This gave him an idea, about pendulum motion. In the cellar of his house he affixed a five kilogram bob onto a steel thread, and set it swinging. At 2 a.m. on Wednesday, 8th of January 1851, he observed the plane of the pendulum gradually turning 'in the direction of the diurnal movement of the celestial sphere.'(Recueil, p.378) Let us surmise that he started the experiment at 1 a.m.

The experiment was repeated in the Paris Observatory with a much longer pendulum, then a church in Paris was used, with a large iron ball and a steel wire more than two hundred feet long under its dome. Some sand was sprinkled on the church floor, and the end of the pendulum had a spike that scored a mark to show its direction. The citizens of Paris could witness the very motion of the Earth as the plane of the pendulum swing moved in the sand. For the first time, here was direct, physical evidence that the Earth rotated.

Figure 9.11: Foucault's Pendulum. (*Mary Evans Picture Library*)

Thus, Jean Foucault demonstrated that the plane of a pendulum's motion will remain fixed with respect to the stars. The stars provide an 'inertial frame of reference.' Foucault showed in effect that a pendulum at the North Pole would stay in the same plane against

EUREKA

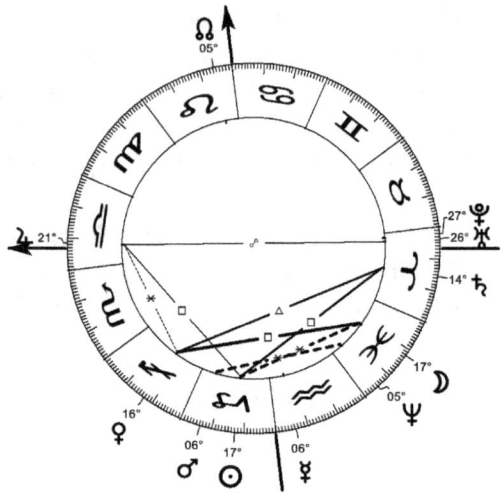

Figure 9.12: Event Chart for Foucault's Pendulum, 8 Jan 1851 01:00hrs LMT, Paris, France. (5° orbs used.)

the stars, with the earth in effect revolving around underneath it. A chart for that time is shown (see figure 9.12). It seems to express a principle of gravity, with planets clustered around the lowest point of the chart, and with a strong emphasis on the line of the horizon.

Telephone

1876, 10 MAR   NO URANUS ASPECTS

An early 'telephone' was invented by Philipp Reis of Frankfurt. On October 26, 1861, Reis gave a demonstration lecture on 'Telephony by means of the Galvanic Current' before the Frankfurt Physical Association. He there demonstrated his 'telephone', able 'to reproduce the sounds of different instruments, and even, to a certain degree, the human voice ... I may add that we are, no doubt, still a long way from the practical use of the telephone.' The response of his colleagues was not overwhelming, but as a result he made and sold a dozen 'telephone' sets. Soon after that Reis died aged just 40, first losing his voice by a cruel twist of fate. 'I have given a great invention to the world,' he whispered to a colleague shortly before his death, 'I must leave it to others to carry it further'.[6]

Chapter 9. When Inventions Worked

Alexander Graham Bell was studying at Edinburgh University in 1862/3 and took a keen interest in one of the Reis telephones which arrived there. On a visit to Professor Wheatstone in London, one of the inventors of the telegraph, he discussed how to represent different frequencies by a fluctuating electrical current. He then emigrated to Boston, Massachusetts, finding employment in a special school for teaching the deaf and dumb, where he became absorbed in the study of sound and how it could be transmitted.

On June 2nd, 1875, Bell and his assistant Watson transmitted the sound of a vibrating reed along a wire. He realised that the frequency of sound could be carried as an electrical pulse, and so it should be feasible for a human voice to be carried along a wire likewise. 'To him [i.e. Bell] that sound meant one thing. The secret of making wire talk had been revealed. Bell's instant recognition of the twanging of the wire was the result of long years of patient investigation'.[7] Another biographer affirmed of that date, 'The telephone was born, and its wise father knew his child, that sultry afternoon.'[8] The chapter, 'The Telephone is born' described the events of that day.

On the evening of March 10th the next year, Bell's assistant Watson was startled to hear the words 'Come here, Watson, I want you' when Bell was in the next room, and realised that the message had been transmitted by wire. That night Bell wrote to his mother, 'March 10, 1876 — This is a great day with me. I feel that I have at last struck the solution to a great problem and the day is coming when the telegraph wires will be laid on to houses just like water or gas, and friends converse with each other without leaving home.' He managed to beat another competitor to the patent office by a few hours.

A fierce court battle ensued which Bell eventually won. A mere two years later he was demonstrating his invention to Queen Victoria. Rich and famous, shortly before his death Bell remarked, 'I am sure that I should never have invented the telephone if I had been an electrician. What electrician would have hit upon so mad an idea?'

The Dictionary of Scientific Biography stated 'on March 10 Bell became the first person ever to transmit speech from one point to another by electrical means' so that date was used. It seems likely that we erred in this, and that the earlier date (with two Uranus aspects) should have been used.[9] [10]

On June 2nd, in the early afternoon, there was a Uranus square Saturn and sextile to the Sun, and Uranus was exactly trining its position in Bell's natal chart (within $\frac{1}{2}$°). By March of the next year, as his achievement was becoming recognised, Bell was into his Saturn-return (3°) and transiting Saturn was conjunct his natal Mercury. His Uranus-trine return was fading out at 4° of orb. No Uranus-aspects were present on that day.

Phonograph

1877, 06 DEC ♂△♀

In July 1877, a thirty-year old Thomas Alva Edison found that he could reproduce recorded sound. At the time he was pondering how to record and automate Morse code signals. He shouted 'Halloo, Halloo!' into the horn of his contraption, and on turning the handle a faint 'Halloo, Halloo!' was heard as it played back. A steel stylus made indents on a paper soaked in paraffin wax, and it was connected up to a telephone transmitter. On notepaper dated '18 July' Edison wrote: 'There is no doubt that I shall be able to store up and reproduce automatically at any future time the human voice perfectly.' The pressure of other work prevented him from doing this right away.

Word began to get around that Edison was planning to make a machine that could speak. Had anyone else announced such a scheme, it would have been laughed at. However, his reputation was then such that, as one observer put it:

> By the simple inhabitants of the region, he was regarded with a kind of uncanny fascination, somewhat similar to that expressed by Dr Faustus of old, and no feat, however startling, would have been considered too great for his achievements.[11]

The *Scientific American* commented on Edison's intention, then Nature repeated for its readers that Mr Edison was trying 'to make the telephone record the sound it transmits.' With all these rumours circulating, Edison decided he had better get on with it.

On November 29th, he first sketched his design for a phonograph in his notebook for his assistant to construct. Tinfoil was wrapped

Chapter 9. When Inventions Worked

around a cylinder, with a needle set skimming over it and a receiver wired up to the needle. The records are unclear over the dates — a problem no less present for the electric bulb — and his biographer R. W. Clark found that *only* the notebook of Charles Batchelor, a British technician present, had any record! For December 4th, it read: 'Kreusi made the phonograph today' and for the 6th December: 'Kreusi finished the phonograph today.' These dates are all we have to go on.

On December 4th, Edison announced to his assistants that he was going to build a machine that would speak. They laughed at him, and one wagered a cigar on the matter. His assistant, Kruesi, was given the design of the apparatus, and spent thirty hours without sleep (such were the conditions which Edison expected) to assemble it, without much idea of what it was supposed to be. Edison then spoke into it, 'Mary had a little lamb' and was startled to hear it spoken back to him as the handle was turned. 'Mein Gott in Himmel!' exclaimed Kruesi. Edison received the cigar. 'I was never so taken aback in all my life,' Edison later recalled, adding, 'I didn't have much faith that it would work.'

A day or so later (the date is not recorded), the machine was taken to the *Scientific American* office where it caused a great stir. Startled readers were informed that: 'The machine enquired as to our health, asked how we liked the phonograph, informed us that it was very well ...' The mantle of 'the wizard of Menlo Park' then descended upon Edison. Once he had made a machine talk, people believed he could do anything. Huge crowds came to hear the new wonder. The first musical recording was made the next year.

'I've made a good many devices but this is my baby', Edison told journalists. By one of those synchronies so common in the history of invention, the French physicist, Charles Cross, was exhibiting before the Academy in Paris on April 30th, 1877, a 'speaking apparatus' to record sound. On 10th October, a description of Charles Cross' apparatus, christened 'the phonograph', appeared in 'La Semaine du Clergé'. The academicians took little notice of his invention, and Cross set it aside.

Was the Edison Electric Company perturbed by priority claims made on behalf of Cross? Persons within that company manufactured a spurious date of August 12th, 1877, as the day when Edison gave his design to Kruesi to assemble. Old books all cited this date, later revealed as a deliberate forgery accomplished by the 'Edison Phonograph people', on the 40th anniversary of the phonograph in 1917! The date was written onto an authentic sketch of Edison's, making the August 12th date appear genuine. Presumably they promoted the loss of genuine documents which could reliably have dated the event. Decades later the fraud was detected, the notebooks of Charles Batchelor being presented to the Edison National Historic Site, establishing the December dating.^{12}

A chart for the phonograph has Uranus acting as a focus. It is untimed, so the Moon's position is blurred. Although we have scored only one Uranus aspect, this chart may remind us that, taking wider orbs and including three more aspects than the five Ptolemaic ones, the picture can look different. When Edison had his idea, on July 18th, Uranus was in exact opposition to his natal Sun (18′), then when the phonograph was made in December, Uranus had reached the middle of his Saturn-Neptune conjunction (4°).

Electric Light

1879, 23 OCT ♅⋆☿, ♅☌♃, ♅△♆

In the Autumn of 1878, Edison announced that he was going to try and make a filament electric light. Inventors had been attempting just that for half a century and got nowhere. But, merely from that announcement, illuminating gas stocks tumbled at once in both New York and London. He was 'The Wizard of Menlo Park', at a mere thirty-one years of age, and for him to announce such an intention was enough to cause apprehension around the world that the era of gas illumination might be drawing to its close. Was this the moment in time when world leadership in technology was seen to cross the Atlantic? In such terms did Asimov see it.^{13}

The Edison Electric Light Company was floated with three thousand hundred-dollar shares. Electrical lighting was then only available as arc-lights, where a high voltage sparked continually across carbon electrodes which burnt out quickly, an expensive process unsuitable for indoor use. A British Parliamentary Report declared the new goal to be impossible, that 'It is certainly not going to take the place of gas.' *The Electrical Review* of February 1879 declared,

Chapter 9. When Inventions Worked

'Electric light is an absolute ignis fatuus.'

'The electric light became feasible because of improvements in vacuum technology. For fifty years people had been trying to solve the problem of the carbon burning away by placing it in a vacuum, but none of the vacua were efficient enough, because the presence of minute traces of air made the carbon oxidise. In October 1879, a new pump arrived at Menlo Park from Germany, made for the purpose of investigating the behaviour of electric currents in vacuum tubes, which could attain a pressure of a millionth of an atmosphere. It provided Edison with an almost perfect vacuum.'

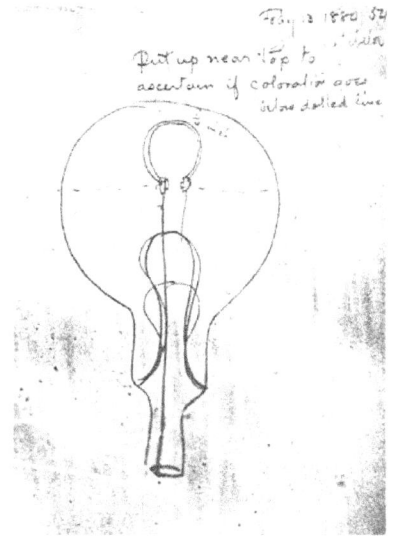

Figure 9.13: Electric Light.

Two hundred notebooks and 40,000 pages chronicled the long, gruelling endeavour. It was said that Edison could never comprehend the limitations of the strength of other men, as his own physical and mental stamina always seemed to be without limit. He could work continuously as long as he wished, and he 'had sleep at his command'.[14] Midnight meals were brought in when the men were working late. Exploding lamps were tried, melting lamps and coiled platinum around a refractory oxide core, with over a thousand tests of earths, minerals and ores. 'My chief difficulty was in constructing a carbon filament, the incandescence of which is the source of light,' Edison recalled. He finally tried ordinary cotton sewing-thread, carbonised in a vacuum. 'So potent was the spell of Edison's inspirational genius, that Mr Batchelor (an assistant) at once yielded to Mr Edison's frantic suggestion that they should make a lamp before they slept, or die in the attempt'.[15] Finally, 'Upon the morning of the 21st' — a mistaken date, as we shall see — the lamp was made, and in the middle of the night their bleary eyes beheld what they had so long desired.

The notebooks kept over those long, sleepless nights were chaotic, leading to a misplacement of the date for this achievement. That generally quoted is the 21st of October when a lamp burnt for 40 hours — while, the *Biographical Dictionary of Scientists* cites the 19th. Edison's careful biographer R. W. Clarke left the moment uncertain, saying merely that 'The breakthrough came at the end of a 5-day session beginning on 16th October', and that 'The experiments of 21st and 22nd October proved that Edison was over the hump.'

The confusion has recently been resolved.[16] Old notebooks from the laboratory showed that: 'October 21, as far as the laboratory record reveals, came to an end without the dramatic success that subsequent accounts of the electric light's invention attributed to it.' For the next day also, it appeared that 'These observations carry no hint of triumph, no inventor's 'eureka!' to set them off, but unquestionably signify the crucial transition of the electric light search ...in the middle of the night — 1.30 by Batchelor's account — the bulb with the simple carbonized thread was put on eighteen cells and kept on ... After number 9 had burned for $13\frac{1}{2}$ hours, more cells were added at 3.00 the following afternoon. It continued to burn for an hour longer ... when it cracked.' In other words, the lamp finally worked early in the morning of the 23rd!

To confirm this date, a letter was sent to one of the authors of this recent study (Associate Professor R. Friedel) who replied:

> The record does show that lamp number 9 of the series Batchelor made on the 22nd was put on the batteries at 1.30 that night, making it the 23rd. You might also note that this lamp burned a total of about 14 hours total ... The disparity between the record of this October 23 lamp and the traditional date of October 21 is a bit bothersome, but not that extraordinary. One of the things we discovered in research for 'Edison's Electric Light' was the degree to which the men at Menlo Park were a bit sloppy in their recollection of events that were later seen to be of historic importance.

Such a view does not increase one's confidence in the dating of the first phonograph to work, which as we saw was given as December 6th purely from one worker's notebook. People came from all corners of the world to see the new wonder, of the Menlo Park laboratory illuminated by hundreds of electric bulbs. It was the apogee of Edison's career.

However, the world's first filament electric light glowed in Newcastle. Its inventor, Joseph Swan, first publicly demonstrated it on January 18th, 1879, after decades of endeavour (mainly in a greenhouse at his home 'Underhill' in Gateshead). However, to quote the *Shell Book of Firsts*, 'Not until 1880 did Swan produce a true filament lamp, employing a carbonized cotton thread similar to that used by Edison, but rather longer lasting. This was the subject of a patent taken out on 27th November 1880, and under which Swan lamps were subsequently manufactured.' When Edison first heard about Swan's achievement he indignantly commenced a lawsuit, but soon matters were resolved amicably with the setting up of the Swan-Edison company. Synchronously enough, the year 1881 saw both Edison's generating station providing electricity for his bulbs, and the British House of Commons being lit by Swan's filament light.

There were five septiles present at Edison's 'moment of illumination', and the septile harmogram peaked a few hours after switching on the lamp. The Menlo Park team had seen many, many lamps light up and flicker out, so only after it had endured for several hours would they have believed that at last a lamp was working. Mars was at the MC, and altogether there were three Mars aspects, expressing the iron resolution displayed by the team in finally making an electric lamp.

X-Rays

1895, 28 DEC

In December 1895, Röntgen discovered X-rays from a cathode ray tube, and within a few weeks, doctors around the world started taking their patients into physics laboratories. The first recorded use of the new rays was a mere few weeks later, on the 28th of December in Vienna when a clinical X-ray of a gunshot wound was made by Franz Exner, on 28th December (*Shell Book of Firsts*). This is surely a record for the swift application of a new discovery to human problems. Many were nervous of the new ray, holding that to see one's own skeleton was a premonition of death. A London firm advertised X-ray proof underclothing.

EUREKA

Radio Station

1897, 10 MAY ♃☌♄, ♃△♂, ♃□♄, ♃□☽

Many had a hand in the development of radio, but there was one who was seen to make it work, who convinced the world by sending Morse signals without wire, and that was Guglielmo Marconi. Half-British and half-Italian, by 1895 the 21-year-old Marconi was convinced that he could apply the great theoretical innovations of Maxwell and Hertz, concerning electromagnetic waves. The Italian government was not interested, so Marconi turned to Britain where he had family connections. Soon the Post Office was lending a sympathetic ear.

At the customs, Marconi was arrested because of all his suspicious-looking equipment, and Mr Preece from the Post Office had to come and secure his release. The British press derided him as an 'Organ grinder without a monkey' because his strange equipment did not seem to be any use. The inventor was offended by such treatment and refused to accept British nationality. In June 1896, Marconi applied for the world's first radio patent, and gave a demonstration on Salisbury Plain. Preece helped him to install the first-ever wireless telegraph station near Cardiff, where they set up a mast 100 feet high on the cliffs for the antenna, and a receiving station was constructed on a small island three miles away in the middle of the Bristol Channel.

On the first day of the trial, signals were transmitted, but they were weak and distorted. Then, to quote Larsen 'On the second day, May 10th, Marconi's method was used for the first time.' There were delays until quite late in the evening, then finally the Morse signal was received from the island. The young man became a world-wide centre of interest. In July of 1897, Marconi formed the Wireless Telegraph and Signal Co. That evening of May 10th was the first time a radio station functioned. We used that date. A new form of communication became evident to all, a year after Marconi entered Britain.

The birth time of this inventor is known, and so roughly is the time for the event, viz. the evening, through which he became the focus of global attention. This is an unusual situation, that we have reliable time-information for both events, the first such on the I-

Chapter 9. When Inventions Worked

moment list.

The two planets associated with radio are traditionally Mercury and Uranus, for communication and electricity. In Marconi's natal chart these are in trine (3°) and Uranus is in square to his Sun (2°). Mars and Venus are conjunct (5°), which appears in this context as an ability to cope with electrical energy, iron and copper being the metals by whose interaction electricity is generated (as we saw earlier with Faraday). Pluto was exactly conjunct Venus, and these three form a trine to Jupiter.

At the time of the event, the Sun was right over his Venus-Pluto-Mars conjunction (within 4′ in the case of Venus), so forming a trine to the natal Jupiter. Also, Jupiter came into trine with the natal Sun, and conjunct the natal Moon. For becoming the spotlight of publicity, Jupiter and the Sun are appropriate planets. The Moon when the experiment began was conjunct to the natal Moon, then in the evening it formed an exact conjunction with Jupiter, and this reached the MC as the experiment succeeded.

Saturn was conjunct Uranus (1°) at the time, which is quite appropriate for a day which saw the establishment of radio. These two were crossing the ascendant, i.e. just rising as the radio message was received successfully, and the Moon was then in square to them both (3°) and also in square to Mercury, as well as being conjunct Jupiter. At the turning-point of a life, when the individual becomes a part of history, one sees a link-up between the two charts, event and natal.

Powered Flight

1903, 17 DEC ♅ ☌ ☉

Wilbur Wright studied the flight of pigeons, wondering how their wing-tilt could stabilise their flight. To test his theories, he and his brother Orville carried a glider plane to some bleak prairie lands off the East Coast of North Carolina, by a village called Kitty Hawk, 800 miles from their home in Ohio. They assembled and tested it on that remote spot, with the tilt angle of each wing separately adjustable. The next year they tried again, but returned home disillusioned. They held to their vision that men would one day fly, but perhaps not in their lifetime.

Wilbur was invited to lecture on the subject to a society of engineers at Chicago. This stimulated him to construct a wind tunnel and test several designs. To test their new information, the brothers returned to Kitty Hawk, and in September and October of 1902 they made more than 700 flights. To quote Ronald Clarke, 'These experiments of the late summer and autumn of 1902 marked a turning-point. Previously, it is clear, the brothers had looked on their flights as an adventurous hobby. Now, out of the future, they saw the beckoning shadow of man piloting himself ... '[17]

Figure 9.14: The Wright brothers' first flight in Kitty Hawk.

In 1903, the Wright brothers sought for an engine light enough for an aeroplane. They ended up constructing their own, and made a propeller, a device hitherto only used in the sea. After waiting weeks for a mild wind steady enough for take-off, on December 15th all seemed ready. Wilbur made a short flight, but rose only 15 feet, then nose-dived. The damage had to be repaired.

The morning of December 17th brought a sudden worsening of the weather. Drab light from leaden skies mottled the stretches of grey sand, and glinted on pools of frozen rainwater. A few gulls dipped over the surf that roared on a nearby beach. Stepping out of their cabin, Wilbur found a strong wind blowing, too strong to fly. For a couple of hours the two huddled around a glowing stove, listening to the wind whistling by. A lone chicken outside the hut provided them with a supply of eggs.

Who would think of trying to fly in such perilous conditions? But, at 10 a.m., the signal flag was run up and the two began laying the track in the sand. It was so cold they had to return to the stove at intervals to thaw out. Four men from a nearby life-saving station braved the cold to watch, and help if necessary. A young boy who made a living catching crabs also turned up. Orville the younger brother was strapped horizontally into the plane, while Wilbur ex-

Chapter 9. When Inventions Worked

horted the crowd to seem enthusiastic and cheer. Above the sound of motor and propellers rose Orville's shout of readiness, and at 10.35 a.m. he flew, or rather hopped. He went 120 feet, at seven miles an hour, slow enough for a photograph to record it. After that they all went into the hut to warm up.

Was this powered flight? Complained one biographer,

> Although Orville's flight of 120 feet in twelve seconds is today universally accepted as the first flight in history, that is only another of the myths that managed to spring up during those long, silent decades following the death of Wilbur ... at Kitty Hawk on December 17 neither brother accepted the first trial as, incontestably, a proper flight. It had simply been too short.[18]

Wilbur took the controls next, and the plane bobbed up and down for 13 seconds, a few feet above the ground. Next it was Orville's turn again, who lasted 15 seconds.

At precisely twelve noon, Wilbur began the fourth trial of the day. It started dipping and rising jerkily as before, but then by 400 feet its path had smoothed out, and continued to bore straight into the wind, at ten miles per hour:

> It was now, at approximately twenty seconds after twelve noon, with the Flyer calmed to level flight, that everyone knew, without need of stopwatches or measurement, that the machine was flying. His progress was undulating, and the flyer ambled along on a more or less even keel, its motor popping confidently, the large propellers biting the cold wind, the transmission chains clanking in their metal-tube guides.[19]

Then all of a sudden, the plane dived and ploughed into the sand and, to the sound of crackling, twisted wood ... the machine bounced to a halt. Wilbur cut the engine and climbed off. The plane keeled over and smashed. He had done it!

At Wilbur's death the *New York Times* stated that the first three attempts were merely 'jumps' and that Wilbur's flight of eight hundred feet in one minute was 'the first real flight by man in an aeroplane' (June 2nd, 1912). A biographer comments, 'Analysis of the

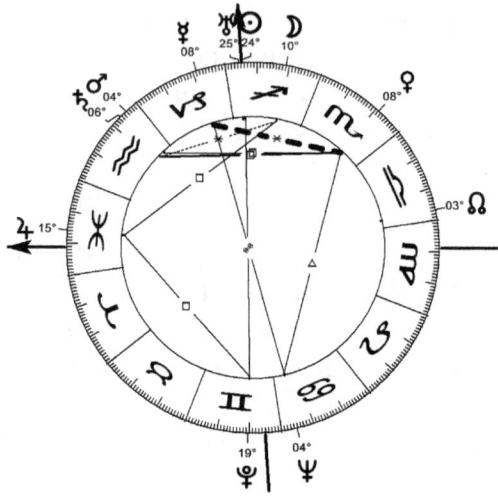

Figure 9.15: Event Chart for Powered Flight, 12 noon precisely EST, Kitty Hawk, North Carolina, USA. (5° orbs used.)

four trials leads inevitably to the conclusion that, while the first three attempts were valuable and heartening, not until the fourth attempt could they be sure that the machine had sustained itself'.[20]

Earlier in 1903, an academic called Newcomb published an article proving that heavier-than-air machines could never fly. He was professor of mathematics and astronomy at Johns Hopkins University, and the vice-president of the National Academy of Sciences. His authoritative article convinced most of his contemporaries of the impossibility of the idea. A professor Langley had been attempting to construct a flying machine, but his funds were cut off after this publication. 'For students and investigators of the Langley type there are more useful employments' intoned the New York Times in a sober editorial.

'A fresh attempt by the Wright brothers in 1904 failed to convince journalists. Such was the effect of Newcomb's article, that several years passed before it was generally believed that the Wright brothers had really flown. British physicist and engineer Lord Kelvin was equally definite: 'I have not the smallest molecule of faith in aerial navigation other than ballooning' he affirmed. In October 4th, 1905, the Wrights achieved the sensational record flight of 25 miles.

Chapter 9. When Inventions Worked

Newcomb only reaffirmed his belief, writing that: 'The demonstration that no possible combination of known substances ... can be united in a practicable machine by which men shall fly long distances through the air, seems to the writer as complete as it is possible for the demonstration of any physical fact to be.' The Wrights had to go to France for a welcome, and the craze for flying really began there.[21][22]

Taking the fourth flight at noon as the I-moment gives the beautiful powered-flight chart with the Sun closely conjunct Uranus on the MC. The Sun at 24° 34′ Sagittarius and conjunct Uranus is right at the hub of the chart: the midpoint of the Moon semisextile (30°) Mercury is at 24° 36′ Sag, and the midpoint of Venus and Saturn, which are in square together (2°) is at 22° 41′ of Sagittarius. Jupiter is just touching the horizon and the Sun conjunct Uranus (1°) reaches the MC. It has an intriguing geometry, and like the balloon charts is focused on the MC.

The Suns in their two charts were linked harmoniously together by a precise trine aspect, within eight minutes of arc, showing the energy of shared endeavour. Orville's chart had Neptune in trine to his Sun and conjunct that of his brother. On the day of the flight, December 17th, 1903, Uranus conjunct the Sun formed a grand trine to these two natal suns. The synastry in the natal charts was activated by the heavens at the moment of flight, on the day when the brothers' dream was actualised.

The degree positions were then:

26° 27′	Leo	Sun	of Orville
26° 19′	Aries	Sun	of Wilbur
24° 35′	Sag	Sun	of flight
25° 45′	Sag	Uranus	of flight
23° 57′	Aries	Neptune	of Orville

Uranus was then conjunct the Galactic Centre at 25° 31′.[23] Some astrologers view 5° Sagittarius as 'the degree of flight', this being the MC of the chart at the time of Orville's first endeavour at 10.35 a.m. (*Mundane Astrology*, p.205.)

Insulin

1922, 11 JAN ᝂᐊᑐ, ᝂᐊᖮ

In the nineteenth-century, diabetes spelt a slow and sure death from which millions perished. Doctors surmised that this mystery illness involved the pancreas, an organ which normally released a hormone called insulin into the bloodstream to control sugar metabolism. Without it, the sugar could not be used and accumulated in the urine. Frederick Banting, a Canadian doctor interested in the problem, approached the University of Toronto to pursue the idea. He was assigned a co-worker, C. Best, and the two of them prepared an extract from an animal pancreas and applied it to a depancreatised dog on 27th December 1921.

Their next step was to apply the remedy to a human subject. Beef pancreas was ground up, extracted with alcohol and prepared and sterilised through various stages. Banting and Best tested it on themselves to make sure it was safe for humans, and then on the afternoon of January 11th, it was administered to a patient. The boy, by then a mere heap of skin and bones, was regarded as doomed unless this new extract worked. His blood sugar did drop somewhat. However, 'no clinical benefit was observed,' and so the doctors in charge would not permit any further injections of Banting and Best's extract.

The extract had to be purified, before it could be of practical use. A visitor to the university named Collip was able to do this. When Banting asked him how he did it, Collip replied, 'I've decided not to tell you.' A cartoon, unfortunately now lost, depicted Banting on top of Collip attempting to throttle him, and the caption was, 'The Discovery of Insulin.' Banting's insulin extract only worked when subject to Collip's purification process. At 11 o'clock on the morning of Monday, January 23rd, the same patient was given a purified extract, which worked.

There are three dates above. We have taken the date of 11th of January, as the date when insulin first worked, in that it *reduced human blood sugar*. No-one except the two inventors saw it that way at the time, if indeed they did. Banting became depressed over the course of events. His biographer tells of 'how desperately unhappy, suspicious, and frightened he had become as the awful pattern of recent weeks had unfolded'.[24] The dog which Banting had initially depancreatised turned out to have some of its pancreas still remaining, so was that perhaps how it was maintaining its blood sugar? A gloss was put on these matters in his Nobel Prize lecture.

Banting was thirty years old when in December he entered Toronto University, and his Uranus-trine-return was synchronising with his Saturn return.

Television

1925, 02 OCT   ☉△♀

'If ever an invention was ahead of its time, this one certainly was', sighed the ageing Herr Nipkow to a journalist. On Christmas Eve, 1883, Nipkow recalled that while 'sitting in my room under the Christmas tree — I found the solution, quite simply and without toil, almost automatically: the basic idea of television'.[25] He invented the 'Nipkow disc', using a selenium cell which reacted according to the light that fell upon it; 'the principle had been discovered: the de-and re-composition of the picture into and from its elements of greater or smaller intensity of light. Between Christmas and New Year, 1883, I was on holiday, and so I had enough time to put my ideas on paper, to make a sketch of my "electric telescope", and have it registered with the patent office. As early as January 1884 I was granted the patent.'

The practical realisation of Nipkow's ideas had to wait for four decades, until a Scotsman, John Logie Baird, assembled it on a sink in Soho. As a small boy, Baird built his own telephone line, but was reprimanded by his father when a coachman became entangled in its wires. He became fascinated by the way selenium was sensitive to light and constructed an optical device. During the war, Baird was declared unfit for service and worked at a Clyde valley power station. The monotony of regular work began to oppress him, until it occurred to him that he could use the electricity to make diamonds. After all, what were diamonds but compressed carbon? He exploded a rod of carbon which had been set in concrete, by electrical means. It didn't make diamonds, but did fuse the power station. After the war, Baird went into business with the 'Baird Undersock Co.' to mass-produce patent socks, which would be 'cool in summer and warm in winter', which he made himself in his Glasgow attic. This was a big success, on account of his use of women parading about in sandwich boards advertising them, the first in Glasgow. After that he produced 'Osmo Boot Polish.'

Baird decided to move to Trinidad, land of 'eternal summer', taking trunks of safety pins with him. He started a jam factory, in a bamboo hut in a village where he was the only white man. He made delicious guava jelly and mango chutney, but swarms of giant ants and mosquitoes overran his jam factory and he contracted malaria, so he had to return home. He could not even sell his insect-ridden West Indian jam. In despair, he recalled the early passion of his youth, and how amplifiers had become available for passing currents through a selenium cell. He wrote to his sister asking her, should he develop his new, patent all-glass razor blade, or television? Stranded in Hastings without money and despite his sister's advice, he set about constructing a television set. He found an electro-motor in some jumble behind an electrician's shop. A hat box was cut out to make his Nipkow disc. Some darning needles, sealing wax, a neon lamp, a biscuit tin, torch batteries, and presto! by the spring of 1924 he was able to transmit the flickering image of a Maltese cross over several yards. A nightmare cobweb of wires and batteries grew across his seaside room.

He moved to an attic flat in Frith Street, in the heart of Soho, where the struggle continued. He found a backer, and finally on October 2nd, 1925, the image of a ventriloquist's doll was transmitted. Baird persuaded an office boy to sit in front of his apparatus. The Nipkow discs whirled and his face duly appeared on the screen.

Some months later on the evening of Tuesday, January 26th, 1926, a number of distinguished gentlemen from the Royal Institution and their wives came over to Frith Street and watched a demonstration. For this demonstration his transmitting equipment was in his attic, and his receiver was downstairs. (On the earlier occasion of the previous year when Baird transmitted a human face, the transmitter and receiver had been adjacent, on the same table). Two members of the Royal Institution went up to the attic, where they were dazzled by a huge lamp and had to move about in front of the revolving Nipkow disc and the selenium cells.

The Times reported that 'the visitors were shown recognisable

Chapter 9. When Inventions Worked

reception of the movements of the dummy head and of a person speaking', adding cautiously that 'it has yet to be seen whether this device will have a practical use.' Readers were told how a beam of light 'crosses the screen so rapidly that the whole image appears simultaneously to the eye.' Overnight, Baird became a celebrity. Two years later he sent a televised picture across the Atlantic by cable, and began using colour television. To introduce television to Britain, Baird had a fight with the British Broadcasting Corporation, which wanted nothing to do with it as radio was quite good enough. 1929, the BBC commenced TV broadcast.

For the time of the historic first demonstration on a Tuesday evening, all we know is that members of the Royal Institution — a society for promoting inventions were accustomed to meet at 9 p.m. on Friday evenings. However, their visit to Frith Street was not an official meeting of that Society.

We have seen how there are two dates associated with the genesis of television, October 2nd, 1925, and January 26th, 1926. The latter is in some respects the main date, and is the one cited by the *BDS*, which said:

> In 1925 he [Baird] achieved the transmission of an image of a recognisable human face and the following year, on 26 January, he gave the world's first demonstration of true television before an audience of about 50 scientists at the Royal Institution, London.

Baird didn't in fact take his equipment to the Royal Institution, but merely invited members to his Soho apartment.

Baird's life reached its nemesis in the first week of November 1936, when the BBC reached the decision that it would go for the American Marconi-EMI electronic system rather than Baird's rotating discs (Pluto was then stationary over his North node). As the BBC held a monopoly there was nothing Baird could do, and he died a tragic figure.

For the I-moment list we took the earlier date, when a face was first transmitted, although the demonstration on January 26th had the better Uranus aspects — three major ones — while the earlier date had just one. The first date saw Saturn transiting his natal Mars. The chart of Glasgow's greatest inventor emphasises the line of his ascendant, with a balance of planets on either side of it. Uranus came down almost to touch his descendant on the first date, then reversed going retrograde so it did not cross his descendant, and then came down again to the very same position on February 27th, a mere day from his main demonstration of television.

Rocket

1926, 16 MAR ♅☌☉, ♅△♄, ♅✶♂

The US national anthem sings of 'the rockets' red glare,' referring to the use of solid-fuel rockets employed in warfare. In 1909, Robert Goddard, Professor of Physics at Clark University, Massachusetts, completed some theoretical studies for a liquid-fuelled rocket. In 1923, he tested such a liquid-fuel rocket engine using gasoline and liquid oxygen. On March 16th, 1926, on a hill near Auburn, Massachusetts, his wife took a picture of him standing next to a four-foot high rocket prior to its launch. At 2.30 p.m., it went up. His diary describes the rocket's slow ascent in a landscape of snow and ice:

> March 17. The first flight with a rocket using liquid propellants was made yesterday at Aunt Effie's farm in Auburn ... Even though the release was pulled, the rocket did not rise at first, but the flame came out, and there was a steady roar. After a number of seconds it rose, slowly until it cleared the frame ... It looked almost magical as it rose, without any appreciable greater noise or flame ...

A granite monument now marks the spot, placed by the American Rocket Society. When German rocket experts were brought to America after the war and were questioned about rocketry, they asked in amazement why American officials did not inquire of Goddard, from whom they had learned, they said, virtually all they knew.

The moment when Goddard's rocket blasted off had some frighteningly exact Mars aspects: Mars sextile Sun, to three minutes of arc, sextile Saturn, 33 minutes of arc, and sextile Uranus, zero minutes of arc. Scarcely less exact are the three Uranus aspects then present: the one as mentioned, then conjunct the Sun to 3 minutes, and trine to Saturn 33 minutes. Speedy Mercury was also forming close aspects, in sextile to Venus (2°), sextile to Jupiter (3°) and square to Pluto (33′). These are all highly appropriate.

Particle Accelerator
1932, 13 APR ♉︎☌☉, ♉︎☌☿, ♉︎□☽, ♉︎□♀

A particle accelerator was constructed jointly by John Cockroft and the Dubliner, E. T. S. Walton, in Cambridge between 1930 and 1932. A high-energy proton beam emerged from their apparatus, with results that made newspaper headlines all over the world.

Cockroft was working at the Cavendish Laboratory under the supervision of Lord Rutherford, the man who had discovered (or perhaps we should say, envisaged) the atomic nucleus. Rutherford was a rumbustuous character known for sayings like, 'any scientific theory worth its salt should be explicable to a barmaid', a view redolent of an era when barmaids had more leisure time. Events at this period of time are of special interest as Pluto had just appeared in the depths of space. An object four billion miles away has relevance to the domain of subatomic particles a mere fraction of an angstrom in diameter ... Does the timing of this invention vindicate such an attitude?

Rutherford waited for three years as the apparatus was gradually assembled. His laboratory had spent five hundred pounds on it, the most it ever spent on a single project. In 1927, Rutherford gave an address to the Royal Society describing how particle accelerators could accelerate to millions of electron volts, but that this was no use because natural radioactivity gave off particles with as much energy or more. The purpose of the apparatus was to seek an entry into the atomic nucleus. Cockroft and Walton built a first version of their accelerator in 1930, then assembled an improved model, ready by April of 1932. A biographer of Rutherford comments, 'Rutherford was impatient while these experiments were being done. He was not interested in the technique, but wanted to know whether the beam could produce any nuclear effects'.[26] Some beams were produced by the accelerator, but it had evidently not yet 'worked'. Then in April, there appeared 'the first disintegrations' of a target placed in front of the accelerator.

Two years after Pluto's appearance, the atom was split. The

dream of the alchemists had been realised, of changing one element into another. Let us quote Walton's own words:

> On Thursday, 13th April, I carried out the usual daily conditioning of the apparatus during which Cockroft was not present in the room. When the voltage and the current of protons reached a reasonably high value, I decided to have a look for scintillations. So I left the control table while the apparatus was running and I crawled over to the hut under the accelerating tube. Immediately I saw scintillations on the screen. I then went back to the control table and switched off the power to the proton source. On returning to the hut, no scintillations could be seen.
>
> After a few more repetitions of this kind of thing, I became convinced that the effect was genuine. Incidentally, these were the first α-particle scintillations I had ever seen and they fitted in with what I had read about them. I phoned Cockroft who came immediately ... He then rang up Rutherford who arrived shortly afterwards. With some difficulty we manoeuvred him into the rather small hut and he had a look at the scintillations. He shouted out instructions such as 'Switch off the proton current', 'Increase the accelerating voltage', etc. ...

The next day, Rutherford returned and swore the two researchers to secrecy. On Saturday, April 16th, at 'about 9 or 10 o'clock that evening' the two of them went round to Rutherford's house, and drew up a letter to Nature. The letter referred to Einstein's $E = mc^2$ equation to explain their results, whereby these latter-day alchemists had managed to change lithium into two alpha-particles. All members of the laboratory were elated by the result, which was the first man-made artificial disintegration of the atom. Shortly after, the discovery was announced to the Royal Society.

There is an apocryphal story that Cockroft and Walton went onto the street exclaiming, 'We've split the atom! We've split the atom!' to anyone who would listen. According to *Reynold's News*, the splitting of the atom heralded 'nothing less than the complete abolition of irksome manual labour and a new era of prosperity for

Chapter 9. When Inventions Worked

all.' A play called 'Wings over Europe' appeared in the West End at the same time as the Cockroft-Walton experiment was making headlines. It concerned the splitting of the atom, and the doomed endeavour to limit its application to peaceful purposes.

A team led by Ernest O. Lawrence was developing the first cyclotron at Berkeley, California. Lawrence learned that he had been anticipated by Cockroft and Walton, and the Berkeley team attempted to replicate this result. The beam of the cyclotron was turned onto a screen of lithium but nothing happened, as the beam was too weak. 'Unfortunately', confessed Lawrence in a letter to Cockroft, 'our beam of protons is not nearly as intense as yours although of higher voltage.' By September of that year, Lawrence finally succeeded in disintegrating lithium. To quote his biographer, 'Berkeley's labour-intensive effort, extending over six months, confirmed the discovery made by a pair of Cambridge physicists in two days'.[27] Particle acceleration devices started to be built in 1930 on both sides of the Atlantic, and they worked two years later. Other remarkable discoveries were also made in that year, so that 1932 was regarded as an *annus mirabilis* by nuclear physicists.

A popular account of these events, in R. W. Clark's *The Greatest Power on Earth*, gives April 13th as the first date cited. 'The first three months of 1932 were used in removing teething troubles, and it was only on 13th April that the first crucial experiments were made'.[28] This date was perceived as the moment when the apparatus first worked, before which it was merely being developed. We therefore take this date as our I-moment.[‡29] Although Cockroft is the better known of the two, it was Walton who set up the apparatus on that

---

‡Hartcup and Allibone in *Cockroft and the Atom* cite the 14th of April as the crucial date, on account of a letter which Walton sent Cockroft years after the event. This begins, 'My memories of what happened in April 1932 are hazy', then gives the 14th as the date when the lithium was disintegrated. Cockroft's notebook (at the archives at Churchill College, Cambridge) quite clearly labels the date as the 13th. Walton's notebook has nothing for that day, but has a write-up of the events dated the 14th. I interpret this as meaning that Walton was too excited on the 13th to write anything up, so that he only filled in his notebook the next day, then years later became confused. The quote by Walton, above, from Oliphant's book *Rutherford: Recollections of the Cambridge Days*, which clearly specified the 13th, was from a letter by Walton to Oliphant, so was presumably penned at an early date before his memory became 'hazy'. The new biography of Lawrence gives the wrong date of the 14th.

day so that it worked and it was he who detected the scintillations.

Taking Walton's chart in relation to the great event, his quite remarkable transits were:

Mars	trine	Mars	1° 44'
Saturn	conjunct	Saturn	1° 27'
Uranus	trine	Uranus	2° 35'
Neptune	sextile	Neptune	0° 28'
Pluto	semisextile	Pluto	0° 41'

He was at that creative period of life when the Saturn-return and Uranus-trine-returns overlap. Pluto had gone round one-twelfth of its orbit, so was in semisextile (30°) return, within 41': all four of the outermost planets were resonating with their natal positions! This alchemic moment displays a unique situation, which we do not elsewhere come across. It gives food for thought as to how the periods of these outer spheres are inter-related for such to be possible. The atom was penetrated, lithium became helium, with four Pluto aspects present in the sky plus four Uranus aspects. Uranus conjunct Mercury/Sun was in a forceful square to Pluto conjunct the Moon.

Radar

1935, 25 FEB ♅☌♂, ♅□♀, ♅⚹♄

Radar has been described as a clear case of simultaneous invention, developing under the cloak of military secrecy in many countries in the years leading up to the Second World War. Powerful short-wave directed transmission had to be achieved before anyone could invent it. Of the secret reports and memoranda that came to light after the war, the most significant — in the view of C. Susskind, author of *Who Invented Radar?* — was that drafted by Sir Robert Watson-Watt in 1935. Indeed, this article — by a Czech, who after all should be fairly impartial — mentions no other name as inventor. Watson-Watt had been asked whether he could design the proverbial death-ray, and replied in the negative, but went on to suggest that a sufficiently directed beam should be feasible to be reflected off a metal target. Its development helped Britain to win the war.

To quote from the article 'Who invented radar?':

Chapter 9. When Inventions Worked

This memorandum stands head and shoulders above all other efforts of the day. Watson-Watt's able reasoning, his grasp of what was feasible, his clear conception of the operational requirements and organisation were all brilliantly vindicated by the success of the British radar network during the Battle of Britain. In this 1940 fight for air supremacy, the slight tactical advantage gained by the Royal Air Force derived largely from the early warning of the approaching Luftwaffe squadrons provided by the radar stations.[30]

In his biography, Sir Robert Watson-Watt describes the first practical demonstration of the phenomenon as being at Daventry on 26th February 1935. 'Late in the afternoon' he set out with a colleague driving from Ditton Park until the masts at Daventry, near Rugby, come into view. They heard the hum of the Hereford bomber as it shuttled back and forth eight miles away, and witnessed the lines it made on a cathode-ray oscilloscope display housed in the back of a Morris van, perhaps just after dark. The Ministry commented: 'It was demonstrated beyond doubt that electromagnetic energy is reflected from the metal components of an aircraft structure and that it can be detected. Whether an aircraft can be accurately located remains to be shown. No-one seeing the demonstration could fail to be hopeful ... In the circumstances the result was much beyond expectation.' That initial experiment was decisive, and that same year the government funded the construction of air-defence radar stations around London.

The Shell Book of Firsts cites the first operation of radar as by a German, Dr Kuhnold. He tested his system in Kiel Harbour on 20th March 1934, when he received echo signals from a battleship at 600 yards distance; later tests picked up signals from a ship seven miles away. The *BDS*, however, cites Watson-Watt as its inventor.

Bearing in mind that flying after dark could then be perilous, we may surmise that the plane was flying until 5 p.m., or so. That would give Mercury on the descendant, with a grand trine in water signs, and a T-square of Mars opposition Uranus in square to Pluto, appropriate for a military invention.

Helicopter

1939, 14 SEP   ♂★☉, ♂★♀, ♂★♆

At the age of 12, Russian-born Igor Sikorsky built a rubber-band powered helicopter which could fly. Later, he became interested in the works of Leonardo da Vinci, and studied da Vinci's well-known design for a helicopter. In 1906, aged 19, Sikorsky visited France and met Wilbur Wright, who stimulated his interest in flight. The following year, he built his first helicopter, but it refused to fly. He established his name as an aviation engineer and emigrated to America. In 1939, he returned to his early interest in designing helicopters, and on the 14th September he piloted one as it rose into the air. The *BDS* gives this date for a helicopter working. Sikorsky's design established the modern helicopter configuration with a single main rotor and a small tail rotor, and rendered obsolete the various earlier multiple-propeller designs. (The first helicopter to achieve free flight had been a twin-rotor machine designed by French cycle-dealer Paul Cornu, test flown at Lisieux on 13th November 1907. It was able to rise six feet in the air.)

Penicillin

1940, 25 MAY   ♂△☽, ♂△♆

In 1928, Alexander Fleming discovered a mould which produced an antibiotic, the 'miracle-drug' penicillin for which the world was waiting. However, 'By 1935, not one person in the world believed in penicillin as a practical aid to medicine'.[31] Fleming had been quite unable to stabilise the substance. This was only achieved in 1940 by the Australian-born pathologist Howard Florey and the German-born Ernst Chain, working at Oxford University. In 1945, Fleming, Florey and Chain shared the Nobel Prize for medicine.

Dr Chain was a Jewish refugee from Nazi Germany, and a most able chemist. He soon found himself working for a Cambridge research laboratory. A couple of years passed, then Dr Howard Florey, who had just been appointed Professor of Pathology at Oxford, came by and recruited Chain to work with him at Oxford. They shared an interest in antibiotics, a field of research regarded as rather barren, and studied Alexander Fleming's paper. Then, by 'sheer luck' Chain

Chapter 9. When Inventions Worked

came across a dry-sounding title in the *British Journal of Experimental Pathology*, 'The Antibacterial Action of Cultures of a Penicillium with special Reference to their use in the Isolation of B. Influenzae.' That had been published in 1929, but strangely enough Chain could find no follow-up to this report. Fleming's penicillin failed to show any marked effects when tried upon patients, and was unstable.

Chain and Florey agreed to study Fleming's new penicillin. Their pathology laboratory happened to have a sample of penicillin mould bequeathed by Fleming at St Mary's Hospital, London. That seemed a good omen. 'In our opinion the purification of penicillin can be carried out easily and rapidly' affirmed Florey and Chain, in their research application. As the War began, a skilled team developed methods of breeding, extracting and testing the mould. Dr Florey realised that the drug had amazing powers, though harmless to animals. Chain later tried to argue that Florey had never been very interested in penicillin and that most of the initiatives had been his, but in fact Florey performed the injection in the crucial experiments to which we now turn.

On the morning of Saturday 25th May 1940, Florey injected eight white mice with a mortal dose of streptococci. An hour later, he gave four of them penicillin. By early evening 'the control animals were showing evident signs of distress.' Florey and colleague stayed up all night to watch, as four 'control' mice died, while three out of the four treated mice remained alive. On Sunday morning, Chain reported, 'It looks like a miracle.' The most powerful drug ever known had been discovered. On the evening of May 25th, it was realised that the new drug was working. In August, *The Lancet* carried their report, to the astonishment of Alexander Fleming, who knew nothing about their work. The first use on a human being was in February 12th, 1941, on an Oxford policeman, and Fleming was present to see his new drug in action.

When Fleming discovered penicillin, Jupiter was transiting his natal Saturn ($1°\,44'$), and this was again the case when he witnessed its first use on a human being: there was an exact Jupiter-Saturn conjunction ($12'$) within three degrees of his Saturn. A Jupiter cycle had passed by, from the conception to the birth of penicillin. Uranus was then conjunct Fleming's natal Jupiter ($1°\,31'$). We may also note that Fleming's Uranus was conjunct Florey's Mercury ($1°$), and in opposition to Chain's Saturn (3°).

Jet Plane

1941, 15 MAY ♅☌☉, ♅☌♄, ♅△♆, ♅△☽, ♅☐♂

In 1928, Frank Whittle published a paper on the use of the gas turbine for aircraft propulsion, as a source of power more efficient than the piston engine. It took him years to convince the Ministry that the jet was a source of power with any potential.

On April 12th, 1937, he became the first person to start up a totally new kind of engine, a turbojet, the basis for all future jet engines. This was during his Saturn return and also his Uranus-trine return, in his thirtieth year. In May of 1941, an aeroplane based on his new design was constructed. Prior to its historic maiden flight on the 15th of that month, it had merely done some taxiing along the runway at an unspecified date.

Figure 9.16: The classic moment, caught by an amateur camera: FW shakes Gerry Sayers's hand after the maiden flight 1941.

As the jet plane was launched, there were seven Uranus aspects in the sky! Five of them were within our 5° orb. There were also four Mars squares: to Jupiter 2°, Uranus 3°, Venus and Sun, an appropriate moment for blast-off. At 7.35 p.m., the Gloster-Meteor left the runway, as the Moon trined Uranus within seven minutes. Its maiden flight soared to 1,000 feet and vanished amidst the clouds. The spectators heard the distant roar of its engine. Shortly after take-off, someone slapped Whittle on the back and said, 'Frank, it flies!'. The flight lasted 17 minutes, a historic moment for British aviation. No-one had filmed it nor had any celebration been planned, as it was all rather unexpected, but an impromptu party was arranged for the evening.^{32}

The new jet plane was held firmly down on the ground right through the War by Government red tape. Government experts saw all sorts of reasons why it couldn't work. A visiting American company became interested, however, and soon there were more Ameri-

Chapter 9. When Inventions Worked

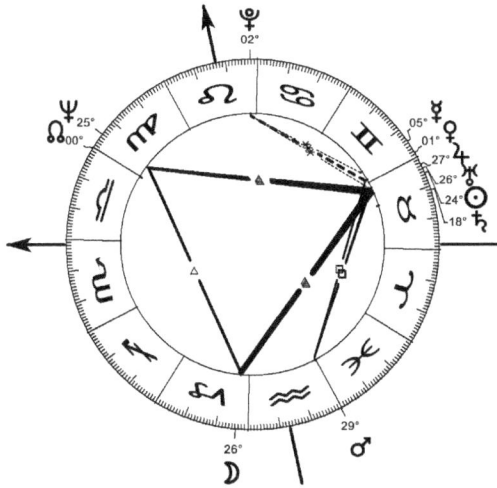

Figure 9.17: Event Chart for Jet Plane, 15 May 1941 19:35hrs BDST, Rugby, UK. (5° orbs used.)

can jets than British, with America using the Gloster Meteor fighter design by the end of the War. Whittle emigrated there. Though the Germans had produced a Messerschmitt jet 262 slightly earlier, it was Whittle's design that became the parent of post-war jet engines.

The close Sun conjunct Uranus (1° 37′) at the time reminds us of the same aspect present in both the Kitty Hawk chart for powered flight (1° 11′) and in Goddard's launch of his rocket (0° 4′). (The helicopter chart had the Sun and Uranus forming a grand trine together with Mars, while the Sputnik chart had an exact Sun sextile Uranus aspect (3′)). Quite a few of the I-moments display grand trines, but none are finer than that for Whittle's jet plane. Let us represent the cluster of six planets centred around Uranus by three pairs of midpoints (figure 9.17).

Starting with the outer pair of

♂□♅
♂□☉
♂□♃
♂□♀

Figure 9.18: Jet Plane, showing Mars aspects.

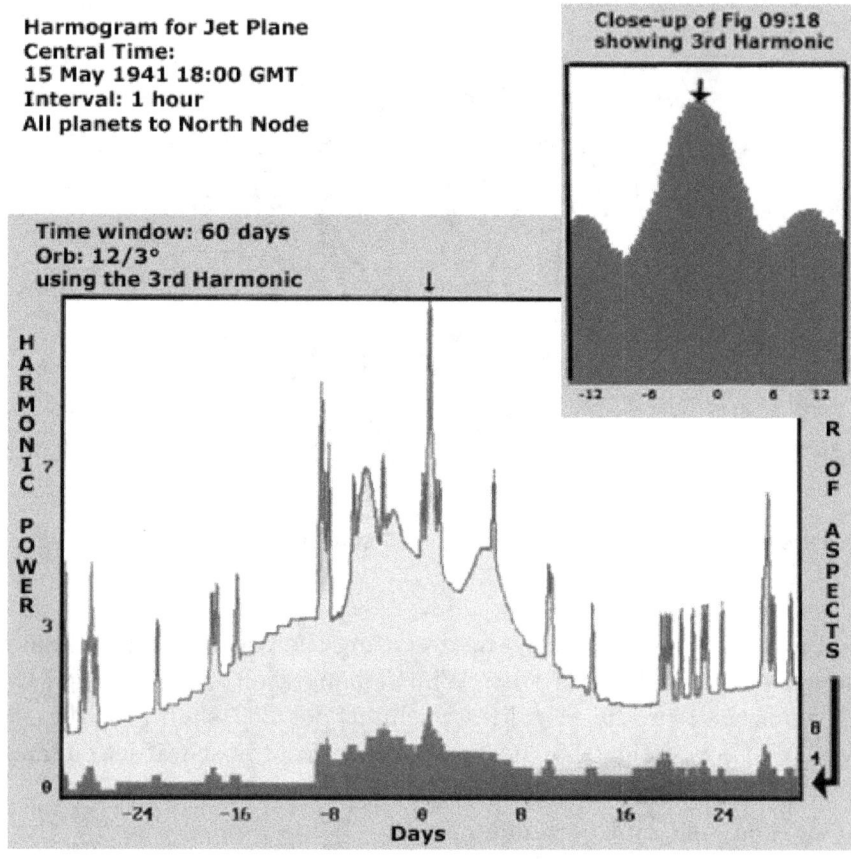

Figure 9.19: Trine power Harmogram for Jet Plane ±30 days with Close-Up.

planets Mercury and Saturn, they are:

Mercury-Saturn at	27° 16′ Taurus
Venus-Sun at	27° 58′ Taurus
Jupiter-Uranus at	26° 47′ Taurus
as were in trine to:	

Neptune at 25° 2′ Virgo and Moon at 26° 13′ Capricorn

Astrologically, the plane took off when it did because the Moon came into this exact trine alignment. As for having this grand trine fall in the three Earth signs, we refrain from comment.

The harmogram (figure 9.19) shows the build-up of trine-power, i.e. of third harmonic aspects, before and after this historic moment.

It shows how over a two-month period, the trine power was strongest on just the day of May 15th, while the close-up shows how, on that day, it reached maximum about half an hour before Whittle's Meteor soared off into the clouds. The strong trines show the vibrant energy of this moment.

Nuclear Reactor
1942, 02 DEC ♂△☽, ♂△♆, ♂⋆♀

On the evening of December 2nd, 1942, Laura Fermi hosted a party for a top-secret department at the University of Chicago. The eminent guests all congratulated her husband as they came in one by one, but no-one would tell her what it was all about. Only two years later did she find out that on that day her husband had masterminded an experiment to unleash the power of the atom. It had been quite a Mars return. 'The Italian Navigator has entered the New World' was the telephone message of early afternoon to announce its success. Wrapped up in the secrecy of the Manhattan project, a Uranium pile went critical.

Blocks of graphite and tons of uranium had been assembled. Graphite slowed down the neutrons, Fermi had calculated, so they would be captured by atomic nuclei. In the morning of December 2nd, the team started gradually pulling out control rods, which adjusted the reaction. The neutron density built up so that the temperature rose. At lunch time they left it and returned at 2 p.m. As the chain reaction developed the group nervously listened to the clicking of counters.

The experiment proceeded by small steps, 'until it was 3.20,' recalled Fermi's wife Laura. Then Fermi turned and said, 'This will do it. Now the pile will chain-react.' A steady nuclear fire was present. A so-called 'suicide squad' of three young men by the top of the pile had buckets of cadmium solution, ready to splash it on to stop the reaction in case anything went wrong. Tension heightened, as the neutron count kept rising. At 3.50 p.m. a smile spread across the face of Fermi, and he ordered the pile to be shutdown, by plunging home the cadmium rods. 'The Greatest Experiment of All Time' was the *Bulletin of Atomic Scientists* title for their report on the matter.

Leo Szilard loitered on the balcony. When it was all over, he recalled how he shook hands with Fermi 'and I said I thought this day would go down as a black day in the history of mankind'.[33] Those two were primarily responsible for the experiment. Szilard dreamed up the notion of a chain reaction, in an E-moment on the streets of London back in 1933, and in the next year Fermi grasped how neutrons could enter the atomic nucleus: there were two eureka-moments presaging this I-moment. All Szilard's dreams and hopes for atomic energy were dashed, as he saw it being used for war.

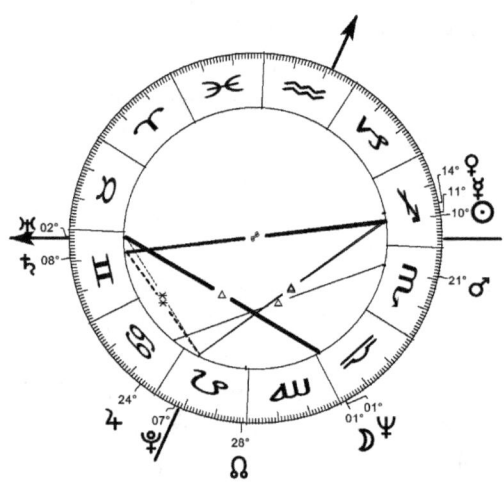

Figure 9.20: Event Chart for Nuclear Power (at culmination), 2 Dec 1942 15:50hrs CST (N.B. War Time not used in lab), just outside Chicago Illinois, USA. (5° orbs used.)

This is *the key* chart for atomic energy. Uranus-conjunct-Saturn was just rising and in opposition to Sun-conjunct-Mercury and Venus. The Moon conjunct Neptune (26′) is in trine to Uranus (54′). The scientists calculated that the new element, later named plutonium, was probably being formed in this pile, as indeed it was, but they had no means of getting to it.[§] Pluto is in trine to the Sun-Mercury

---

[§]Fermi had actually achieved this some years earlier. On May 10th, 1934 in Rome, Fermi transmuted uranium by neutron bombardment, ending his diary for that day with the comment, 'Element 93 may have been found.' Actually he had split the uranium atom, but didn't realise it. *The New York Times* gave a headline 'An Italian Creates a New Element by bombarding Uranium' for that day, when Uranus was exactly conjunct Fermi's natal Moon (20′), and opposite

Chapter 9. When Inventions Worked

conjunction. The Sun-opposition Saturn (1°) is aligned to the two important first-magnitude zodiac stars Aldebaran and Antares, which are 180° apart in the sky as shown: the Bull's Eye and 'Heart of the Scorpion' (see page 293, The Stars of Conflict). Saturn was one arc-minute from Aldebaran.

This powerful opposition was crossing the ascendant-descendant axis just as the day's experiment culminated. (N.B. The military who conducted this experiment were not on War Time as was the rest of Chicago. Thus, to obtain GMT we subtract the normal six hours from the time given, making 15:50 - 6 = 9:50hrs. The co-ordinates were 41N46, 87W37, which is just outside Chicago). Pluto, as 'the bomb in the basement' was right on the IC, somewhat implying that Fermi switched the pile off just in time.

The axis of this chart is exactly linked to the Pluto of the Hiroshima chart, and the Hiroshima Mars is exactly conjunct Saturn of this chart, to 14′, i.e. it is right on the 'nuclear axis' as some call it. The stars are telling us about the military orientation of this project. The Hiroshima bomb was the first uncontrolled uranium fission reaction, whereas the Chicago experiment was a controlled fission reaction.[34]

Fermi's Mars was conjunct the nuclear power Mars within a degree. His Uranus was conjunct Szilard's Saturn (standing at 13°30′ and 11°5′ Sagittarius, respectively), so they were immediately conjunct the Sun-Venus-Mercury cluster at the focus of this chart. Szilard's Pluto-Saturn opposition was well aligned with the major axis of the nuclear energy chart, with his Saturn conjunct the Mercury of the nuclear power chart within an arc-minute, and its Sun to one degree.

It was Mark Lerner who recognised the key importance of the Antares-Aldebaran axis in nuclear energy.[35] This axis is 'remembered', he has argued, by major nuclear accidents, notably those at Chernobyl and Windscale. On October 7th, 1956, a destructive fire occurred at the Windscale nuclear plant, as Saturn at 10°5′ of Sagittarius stood within a couple of arc-minutes of the Sun of the nuclear power chart, and loosely conjunct Antares, 'heart of the Scorpion.'

---

his Mercury ($\frac{1}{2}$°). Later, in October of that year, he came to understand the principle of slowing down neutrons (see chapter 5) but still had no idea about fission, as was later used in the Uranium pile.

Then on April 26th, 1986, the Chernobyl reactor blew its top, one Saturn-cycle later, with Saturn at 8° 38′ of Sagittarius.

Atomic Bomb

1945, 16 JUL   ♅△☽, ♅✶♀

At 5.38 a.m. at Alamagordo Air Base in New Mexico, on July 16th, 1945, the first atomic device was exploded. Two hemispheres of plutonium came together as Mars passed by the baleful star Algol (20′ orb). The week before this saw a total solar eclipse on 16° 57′ of Cancer, which is within a degree of Pluto's position at its discovery. Saturn was conjunct this eclipse at 14° 44′ of Cancer. Pluto at 9° Leo was opposing the Mars of the Pluto-discovery chart.

Weeks later, two hemispheres of uranium came together for the first time over Hiroshima, on 6th August. Saturn was closely conjunct the Pluto of the Pluto-discovery chart, joined in this exact position by the Moon. This is also the degree of the solar eclipse a month earlier. Jupiter was within four degrees of its position at the chain-reaction E-moment of Leo Szilard. Uranus, conjunct (7°) Mars, was crossing the Hiroshima MC over the minutes when the Bomb was dropped.

Mars at 9° 12′ of Gemini was transiting the 'nuclear axis', so was conjunct the Saturn-position in the nuclear power chart (see figure 9.20), within about fifteen arc-minutes. Pluto, in an exact sextile to Mars over Hiroshima, formed an exact right-angled triangle with the 'nuclear axis' Antares-Aldebaran: sextile to its Saturn-conjunct-Aldebaran (1°) and trine its Sun-conjunct-Antares (within 4′).

Transistor

1947, 16 DEC   ♅☍☉, ♅✶♄

Bell Laboratories were the research arm of the world's largest telephone company. A team, under the leadership of William Shockley, was seeking an alternative to the thermionic valve, as these were fragile, bulky and took a while to warm up. The team was looking at semiconductors as an alternative.

What Shockley recalled as the 'breakthrough observation' came on 17th November 1947. They were seeking a means of amplifying current, when they began to comprehend the motion of 'holes' in their semiconductor, i.e. positive charges which moved about. 'This new finding was electrifying,' recalled Shockley, and his notebook records on the subject date from this moment. After that, followed the 'magic month' when so many significant insights were achieved.

The experimentalist of this team was Brattain, and on the 16th December he made an apparatus of two gold wires stuck into a small bit of germanium, which was the first transistor. The point-contact transistor was 'born' on that date. The events of that day were described by John Bardeen in his Nobel lecture of 1956:

> This experiment suggested that holes were flowing into the germanium surface from the gold spot, and that the holes introduced in this way flowed into the point contact to enhance the reverse effect. This was the first indication of the transistor effect ...

Gold was evaporated onto a wedge of germanium, and close point-contacts were thereby achieved. 'Success was achieved in the first trial; the point-contact transistor was born.' Amplification of electricity by means of the three-way transistor was achieved. Recalled Shockley: 'If "conception" is taken as the definition of "invention", then 15th December is confirmed both by my notebook research and by the Patent Department's records — the first observation of amplification was one day later.'

The Reader's Digest *'Inventions that Changed the World'* incorrectly cited 24th December as the date for this new invention. This mistaken date somehow became more widely known than the true date. This meant, to quote William Shockley:

> The failure to correctly date the birth of the transistor had pleasant consequences. It caused anniversaries of the transistor to be held in a warm Christmas Eve atmosphere.[36]

On 24th December, there was a public demonstration for the members of the Bell Corporation and the Press, when a human voice was heard to be amplified. As a public witnessing of the event, it seemed to be in the great American tradition of Alexander Bell calling out 'come here, Watson' when his telephone system first worked.

This small object, destined to grow so much smaller, failed to impress the media, which mentioned it only briefly. In 1956, the three inventors, Bardeen, Shockley and Brattain, shared a Nobel Prize. They found that silicon would work instead of germanium.[37]

Use of gold leads us to seek for solar aspects. There was indeed a Sun-Uranus opposition, reaching exactitude *on that day*, the 16th, plus a Sun-Saturn trine, of orb 1° 19', so that Saturn was in sextile to Uranus. These are symbolically the most relevant planets, and for an apparatus which was in essence a three-way junction this configuration seems appropriate.

Computer

1948, 21 JUN ♉︎♂☉, ♉︎♂☽, ♉︎♂♃

Calculating engines had existed since Pascal, but the computer in the modern sense of a programmable electronic device appeared in the mid-1940s. According to the Shell Book of Firsts, 'The first general-purpose [i.e. capable of being programmed] all-electronic computer was the ENIAC (Electronic Numerical Integrator and Computer), developed for the US Army ... and completed in February 1946.'

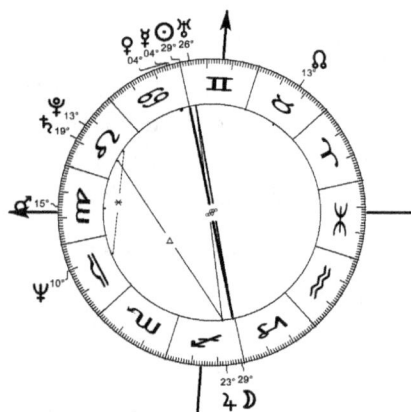

Figure 9.21: Event Chart for Digital Computer Program, 21 Jun 1948 12:00hrs BST (just before lunch), Manchester, UK. (5° orbs used.)

However, before that, the world's first programmable electronic digital computer was Britain's top-secret 'Colossus' computer, designed by Tommy Flowers MBE, and built by him and his team at Dollis Hill, becoming operational by Dec 1943. It was then dismantled and set up in Bletchley Park¶ The Mk II version was operational in June of 1944, and helped crack the encrypted German submarine messages. It had to be programmed manually, not having any internally-stored electronic digital programs. It was not general-purpose, but programmed for specific decryption tasks. The general-purpose (i.e. *fully*-programmable) ENIAC was built from the Colossus designs, given to the US by the UK as part-payment for war loans.

The first *stored-program* computer was built at Manchester University. Its first program was run on 21st June 1948, just before lunch. Its inventor, Frederick C. Williams, was an electrical and electronics engineer, holding the Chair of Electro-technics at Manchester University. A biography of Alan Turing described how, on that day of 21st June 1948, the Manchester team

> successfully ran the first program on the first working stored program electronic digital computer in the world.[38]

While there was no definitive 'first computer', there was a distinct moment when the first stored program was fed into an electronic digital computer. We took that moment. (See figure 9.21.)

The genius of Alan Turing was behind the early development of computers, pioneering computer algorithms and designing 'the Bombe', an electro-mechanical device used to decrypt the Enigma codes during Turing's Uranus-trine and Saturn returns. His life became constrained by bureaucratic security measures after the war in consequence, made worse by 'the love that dare not speak its name', his being a homosexual. He was never recognised or given proper credit, nor did he manage to be around when the first computers started to work after the war, and took his life.‖

¶See the table in Wikipedia, 'some early digital computers: http://en.wikipedia.org/wiki/Colossus_computer.

‖Indeed, none of the pioneers at Bletchley Park received recognition for their work until very late in their lives, as their work was classified 'Top Secret' for a long time. For Tommy Flowers, the scant recognition came too late in his life,

203

EUREKA

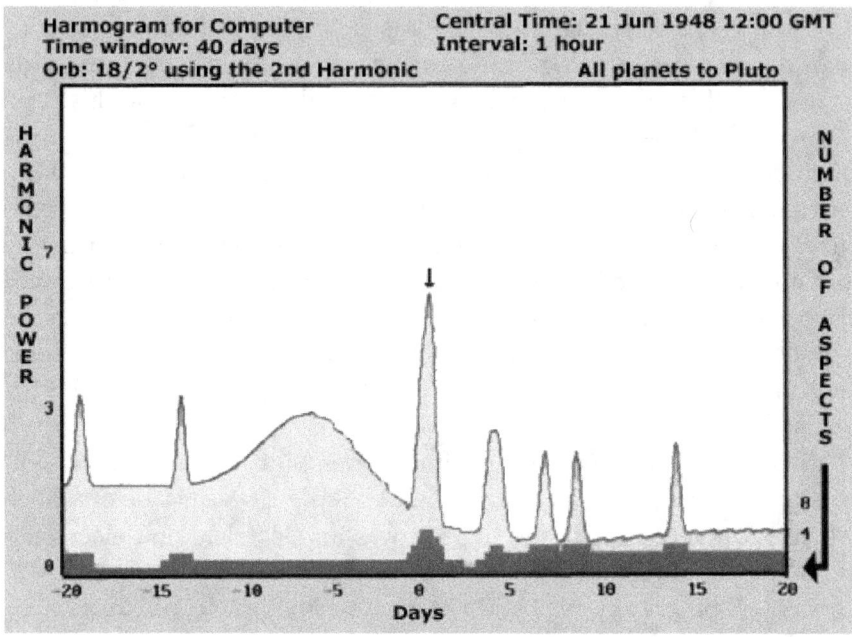

Figure 9.22: Harmogram for the Manchester Computer running its first program, with the 2nd Harmonic, i.e. opposition aspects centred on noon, ±20 days.

Turing's Mercury was conjunct the Mercury of the chart for the Manchester computer's first run. There was a Full Moon on that day, with Uranus conjunct the Sun, all lined up with the Galactic Centre. (Uranus was opposing the Galactic Centre within half a degree). Agile Mercury is at the heart of this high-energy chart, conjunct Uranus (though wide, at 8°) so that the midpoint is right over the Sun-Moon axis. It is in quintile to Mars (10′) and has a strong semisquare to Saturn (16′), and is within minutes of Venus, appropriate for all the conduction of electricity by copper wire on that historic day.

The transits of F. C. William's chart to the Manchester computer chart are strong, though his time of birth is unknown. The Sun was conjunct his natal Sun, Moon, Mercury and Pluto (4°, 1°, 5° and

---

Sadly, this creator of the World's first computer still remains relatively unknown to this day, along with Bill Tutte, who cracked the enemy's algorithm. To compound matters, Alan Turing is often wrongly given credit for work on Colossus, in which he had no input whatsoever - a point firmly asserted by Tommy.

2°, respectively), in other words, he was born on the day of a New Moon, and his computer worked near to his birthday. Taking a noon chart, the Sun was conjunct three other planets, so this is quite a powerful synastry and one can only regret that Williams has left us no account of how he felt on that day. Mercury of the I-moment chart was also conjunct his Sun and Moon (16′ and 1°, respectively), and Uranus was conjunct his Sun, Moon and Mercury. Saturn was conjunct his Venus (26′), Saturn being important for moments of scientific discovery.

A computer works with binary logic, which means on/off, right/wrong, positive/negative, yes/no, true/false. Accordingly, a harmogram for the 'second harmonic', in other words the 180° opposition aspects, was created for the weeks before and after the first program was fed into the Manchester computer (figure 9.22). This harmogram helps us to appreciate why it functioned on just that day. The strong 'second harmonic' power shows in the chart as multiple close oppositions. Usually oppositions signify stress or difficulty, but here they are expressing the very principle of duality! All its diode valves had to be in one of two conditions — on or off. The massive peak of the harmogram was in the afternoon, several hours after the program had been run. Surely, Alan Turing would have appreciated the marvellous logic of this diagram.[39] [40]

Atomic Clock
1948, 12 AUG   ♅★♄

The quartz watches we use today were first discussed in the year 1948. On April 30th, Dr Harold Lyons presented the ideas to the American Physical Society, Washington. According to their report:

> Dr Lyons presented a discussion of atomic clocks and frequency standards using spectroscopic methods in his paper. This was the first proposal of a clock using the method of stabilising a complete quartz-crystal oscillator-frequency multiplier chain.

Then, 'On August 12th, 1948, the world's first atomic clock was given its initial run'.[41] The oscillation of ammonia molecules was somehow linked to a quartz crystal. The oscillation at a megahertz frequency showed as a single spectral line. A test on December 31st showed that the method was viable. Before this design was completed, the team started work on a caesium beam clock that would have a greater accuracy. The work at the Washington Bureau of Standards stimulated Britain's National Physical Laboratory, where the official switch-over to caesium beam time was accomplished in 1955. A press announcement was made on January 6th, 1949, and a week later Dr Lyons was interviewed about his discovery, with the beeping of the atomic clock put on the air.

Thermonuclear Device

1951, 01 NOV ♂☌♂, ♂□☽, ♂□♄, ♂□♇, ♂⚹♆

At the end of the War, many of the scientists who had been involved in the Manhattan Project decided to return to civilian life and forget about building bigger bombs, but Edward Teller was not amongst these. He nursed the notion that an even bigger bomb was somehow necessary. For some years this appeared to be impossible, and according to Jungk, Niels Bohr pursued research in this area in order to demonstrate that it was not feasible.[42]

For a while it looked as if Mother Nature might be successful in hanging onto her secret, the power that lights the stars. Edward Teller was assisted in the complicated calculations necessary by a new calculating device which Von Neumann at Princeton university constructed, replacing the ENIAC computer. This one was called 'Mathematical Analyser, Numerical Integrator and Computer', or MANIAC for short. Teller won funding for a new weapons laboratory, with himself as boss. Teller is generally called the father of the H-bomb, though Hans Bethe commented, 'I used to say that Ulam was the father of the H-bomb, and Edward the mother, because he carried the baby for quite a while'.

A crisis arose at the end of 1950, when Teller became desperate and there seemed no way of making his super-bomb work. By February 1951, anger was developing between the two persons working on the problem, Edward Teller and Stanislaw Ulam, and during this strain a sudden resolution occurred, 'the Teller-Ulam configuration', which became top-secret, whereby tritium would become lithium by nuclear fusion.

Chapter 9. When Inventions Worked

The actual date when the first thermonuclear device exploded has been shrouded in official secrecy. It was the last I-moment which we located, because early books could not date the event. Then, Richard Rhodes' 1986 opus *The Making of the Atomic Bomb*, using recently declassified documents, concluded that 'Mike 1, the first true thermonuclear bomb, [was] tested at Eniwetok in the Marshall Islands on November 1st, 1952'.[43]

Robert Jungk's *Brighter than a Thousand Suns* conjectured that it went off 'Before Dawn' in the Marshall Islands, which may be mistaken. A letter to the Ministry of Defence elicited the reply that their archives had no data for the time of this event, but that I should try the Department of Energy at Washington, which gave the same reply, and suggested approaching the Nevada Department of Energy. They were more helpful, replying that 'the world's first thermonuclear detonation' on Eluklab island was at 0714.59.4 (Marshall Island Time Zone) — however they would not specify what time zone the military were using.

The mushroom cloud spread over sixty miles in half an hour, from a blast a thousand times that of the Hiroshima bomb. It 'vaporised the island of Elugelab.' The Bulletin of the Atomic Scientists report on the event[44] gave the detonation as 7.15 local time, still leaving the problem of what was local time. Normally the time zone for the Marshall Islands is -11hrs, however a zone of -12hrs may have here been used.[45] Dawn at Eluklab would have been at 7.05 a.m. on a -12hr timezone, so the bomb would have outshone the rising Sun. Teller sent off a telegram to Los Alamos, which just said: 'It's a boy.'

It was the view of Arthur Koestler that the most important date for the human race was August 6th, 1945, for since that day at Hiroshima, 'mankind as a whole has had to live with the prospect of its extinction as a *species*'.[46] But, the destructive capability of a fission bomb is limited, and the spectre of global annihilation only arrived with the thermonuclear device.

Uranus, at 18° Cancer, was at Pluto's discovery-position (17° 46' of Cancer), while Venus had just transited over Antares, red star of the 'nuclear axis.' Teller was born during a Uranus-Neptune opposition (18') in 1908, when the nodal axis was also aligned with it to half a minute. The first thermonuclear explosion took place as Mars transited that axis in his chart, conjunct his Uranus within a degree.

207

His Uranus-opposition return was then separating at 5 degrees.

Sputnik

1957, 04 OCT ♅△♄, ♅⋆☉, ♅⋆♂, ♅⋆♃

The Soviets announced in August 1957 that they had successfully tested an inter-continental ballistic missile, but many US scientists expressed disbelief. Then, the Soviets launched Sputnik 1, the first man-made satellite, in October of that year, sending a shock wave of demoralisation across America. Television stations interrupted their programmes to listen to its eerie beep-beep as it passed overhead. The Soviets had achieved the difficult technical task of boosting a rocket cone above the atmosphere so that it moved parallel to the earth's surface, counterbalancing gravity, for which sophisticated rocket guidance systems were needed. The next month a dog was placed in orbit, in Sputnik II. President Eisenhower summoned ex-Nazi rocket expert Werner von Braun and demanded a swift response. A crash programme was initiated, placing a US satellite in orbit four months later. The Space Age had begun. The Russian Sputnik was launched at 9 a.m. GMT, according to The Times, from Georgia just North of the Black Sea.

Hovercraft

1959, 30 MAY ♅⋆☉

In 1950, Christopher Cockerell decided to try and improve a boat's performance. It occurred to him that if he could make the skin of his craft behave like a skin of air, i.e. introduce a film of air between hull and water, then skin friction would be negligible. He conducted experiments with a hair dryer, some tin cans and kitchen scales. A working model of a hovercraft glided across a Whitehall carpet in 1955, and was put on a Secret List so that nothing happened to it.

In December of that year, Cockerell filed the first hovercraft patent. He could do little more, as ship-builders said it was an aircraft and aircraft manufacturers said it was a ship, so no-one would fund it. In 1958, the Secret file became de-classified and the National Research Development Corporation decided to back it. The next year, a craft was built to travel on water over a cushion of air by Saunders-Roe, and launched at Cowes on the Isle of Wight on May 30th. Its first public appearance on June 11th caused a sensation, as it lifted off on a ramp and glided down into the water. That summer, the inventor sailed across the Channel on it.

Laser
1960, 15 MAY

In August 1959, Theodore Maiman decided to begin work on the laser — something that many laboratories were then attempting. Working at the Hughes Research Laboratories in Malibu, he had been asked to research an 'X-ray Maser', as the apparatus was then called. He wasn't happy about the way this required liquid helium temperatures, nor the enormous magnet it required, which was very cumbersome. In 1958, the physicist Charles Townes had published a suggestion about stimulated emission for excited atomic energy states. Townes was concerned with microwave amplification, and used a gas, ammonia. Maiman decided to begin with a ruby crystal, because he believed that ruby was robust and had a simple energy-level scheme. He determined that ruby had a much higher quantum efficiency than had been hitherto supposed. He required a very bright quartz lamp to energise the crystal, and 'de-populate its ground state'. He used a pulsed mode to achieve the necessary power, and could avoid the need for cryogenic cooling.

His calculations revealed that, if it worked at all, the ruby at room temperature should emit visible light. Next, he checked the absorption coefficient along the ruby crystal axis using a parallel plate resonator, then decided to place the ruby crystal right inside the flash lamp, after evaporating some silver onto one end of the ruby surface for reflectance.

No-one had ever produced coherent light before, and one prominent physicist was proclaiming that it was impossible. Another stated that ruby couldn't work, due to its 'ground state' properties, adding, 'Nobody knows what form the laser will take,' which was true enough. Most laboratories in the race were using alkali vapour, and that was far from pleasant to have around a laboratory. Maiman was visited by one scientist, about to give a course at the University of Michigan on why a laser could not be made.

Stress built up at Hughes Laboratories where Maiman was working, with questions of funding for so doubtful an enterprise. No-one had asked Maiman to make a laser. Maiman found the going an uphill battle, and became stubborn: 'I knew there was something there, and I was determined to follow through on it. I had a sixth sense that it was going to work,' he confessed, in an interview. He did not describe his reaction when, on 10.30 a.m. (Hughes Laboratory document) on May 15th, he beheld a beam of red light emerge from his silver-painted ruby crystal — a million times brighter than the Sun, or so it is said. But then, that is always the way with scientists, of hiding away their own subjective experience. Maiman did add, 'Of course, all hell broke loose after we got the laser working, but that's another story'.[47] It was light of a single wavelength, parallel and in phase, i.e. coherent.

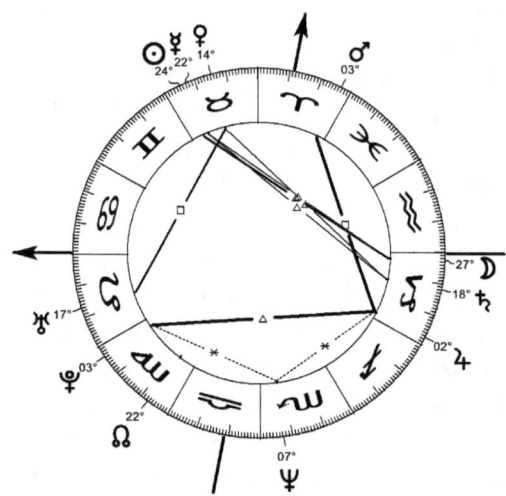

Figure 9.23: Event Chart for Laser Beam, 15 May 1960 10:30hrs PDT, Malibu, California, USA. (5° orbs used.)

The chart of the first laser beam is shown. It has a mandala-like quality, showing the flow of coherent energy. It shows the minute when something emerged perhaps not seen since the first day of Creation — coherent light! Uranus appears as the most heavily aspected planet, if one uses orbs somewhat wider than the 5° we have scored, plus other smaller aspects such as the sesquiquadrate. For this idea, indeed for this memorable chart, one is indebted to the late Ananda Bagley director of Electric Ephemeris.

The 'lunar' metal silver, most reflective of metals, is shown in this chart as literally being at one end of the beam, as the Moon on the descendant. A telephone call to Mr Maiman's secretary obtained the hour of his birth, from which one can see an awesome Mars-Uranus linkup. The beam was red, after all. His natal Mars was right over the Uranus of the Laser beam chart (4°), and his natal Uranus (exactly conjunct Jupiter (10′), auspicious for an inventor), was conjunct the Mars of the laser chart, to within 2′! No wonder he became stubborn. No-one else was going to cheat him of this victory. For a scientific invention, his natal Sun was opposite Saturn to within 7′, also his Mercury was conjunct the ascendant of the laser chart, within 1°. Having such a pair of exactly-timed charts, natal and event, is of great value.

The whole emphasis of this chart falls in the three earth-signs of the zodiac — for a moment which is the quintessence of pure fire! (Sidereally, they would fall in the fire-constellations).

Hologram

1963, DEC   ♅☌♀, ♅△♀, ♅□☉

The hologram used the newly-invented laser beam to produce an 'image' which appeared as a formless blur, but was a diffraction pattern. The laser beam could then resolve it into an image. One could break the diffraction pattern in two and still make a whole image from any fragment, but it would be less distinct. The whole image was somehow contained in every part.

The two research engineers responsible, Dr Emmett Leith and Juris Upatnieks, at the University of Michigan, had found a way of stamping the phase of light beams onto a surface. The light phase was converted into intensity variations by interference, which was then recorded on film. The reconstruction process redirected the rays to make an image. It was based on an earlier theoretical development by Dr Dennis Gabor, an English physics professor at Imperial College, London.

In 1960, Leith and Upatnieks duplicated Gabor's original work, and next year announced to the Optical Society of America their preliminary results. Soon after, the laser became available, permitting them to look at three-dimensional objects. 'Gabor's brilliant conception of optical holography in 1947 lay almost abandoned until Leith initiated in 1963 what became a massive worldwide development effort on it'.[48] It was the three-dimensional possibilities of the hologram that awoke wonder. As early as 1949, Gabor had written that 'the photograph contains the total information required for constructing the object, which can be two-dimensional or three-dimensional,' but his ideas remained theoretical until the laser became available. Gabor had the idea in a fine eureka moment while sitting at his Rugby tennis club one fine day over Easter 1947, but the date is lost. 'In late November or early December 1963', wrote Leith (to the writer) 'we successfully made our first hologram of 3-dimensional, diffusely reflecting objects.'

On December 5th, 1963, the New York papers received a press release about an optical system which could form images without using a lens. An account of theirs captures the excitement of that time:

> For our first attempt the best object scene we could produce was a pile of junk retrieved from odd corners of the laboratory. We were nonetheless quite excited with our first successful hologram. We anticipated seeing a most amazing image, one unlike the world had ever seen before. We knew it would be truly an optical recreation of the original object, fully 3-D, with complete parallax and all of the other properties of real life objects. The basic theory of holography implied all this. We could hardly wait until the development process was finished so we could put the hologram in a coherent light and thereby witness a phenomenon never before seen.

The first one to be made was too small for the eye to appreciate the effect, and the date for a larger one, when the 3-D effect appeared, was the 4th of December as cited in their press release.[49]

Chapter 9. When Inventions Worked

Superconductor
1987, 29 JAN ♀☌♅, ♀□♃, ♀⚹☿

Chinese physicist Ching-Wu Chu made electricity flow with zero resistance, through a special substance he had made. He made this 'superconductor' out of a copper ceramic. Before this, superconductors had only operated close to absolute zero. To quote from a recent book on the subject:[50]

> In January 1987, Professor Paul Chu of the University of Houston made a major scientific breakthrough that was to reverberate around the world ... For many years superconductors, materials that conduct electricity with virtually no resistance, have been known about and available for commercial use — but at operating temperatures too low to make their wider application economic. Paul Chu's achievement was to shatter all previous records and produce a superconductor at workable temperatures ... His discovery is to physics what the double helix was to biology, and its full implications are yet to be realised. When Chu mixed three elements, barium, yttrium and copper, he committed alchemy.

On the 29th of that month, 'At 5.00 p.m. the measurement was made ... All of the Alabama and Houston workers were exultant. It was the discovery of a lifetime.' That was 23:00hrs GMT.

Superconductors are special ceramics made of copper complexes, and copper is the Venus-metal. At the historic moment, there were no less than *five* major Venus aspects taking place: it was conjunct Uranus (2°), conjunct Saturn (5°), trine MC (3°) square Jupiter (20') and sextile Mercury (2°). In addition, three septiles were present, adding a touch of inspiration, the Venus-Moon septile being within 16', and the Saturn-Sun septile within 14'. If we include this septile, there were six significant aspects to Venus, which is a lot. At that moment when Chu 'committed alchemy', the Venus/copper property of conductivity was manifested. Electricity overcame the inertia of matter, to glide through copper without resistance: for which Venus in aspects to Uranus, Saturn and Mercury was appropriate. Uranus, the planet which astrologers associate with electricity, was conjunct the Galactic Centre (1° 30′), loosely conjunct Saturn (7°) and square to Jupiter (2°).

The Moon was in a Sun-Mercury conjunction to around 6° orb, near the midpoint of these two. As Mercury with the two luminaries moved down to touch the horizon, a new property was born in matter.

****

No doubt further I-moments will be located. Perhaps manuscripts will be found, giving the date when Johann Gutenberg in Germany first made his printing press work in 1440, or when Alfred Nobel first made his 'dynamite' (appendix D). Other I-moments, as could be included, are:

Lunar Landing Module

1969, 20 JUL ♉☌♃, ♉⚹☉, ♉⚹♀, ♉⚹♂, ♉⚹♆

'The Eagle has landed.' The words were radioed back to Earth, once the Apollo 11 landing module had settled upon the lunar surface. This spider-like craft could not be tested in Earth-gravity, so this was a genuine first-use — as two astronauts descended, on a pillar of fire, to the surface of another world. 'The Eagle,' the name of the landing module, was a bird traditionally associated with Jupiter, and a Uranus-Jupiter conjunction had chimed that very day, being then (geocentrically) a mere few arc minutes apart. This could have been a moment of grave peril and difficulty, for who really knew whether a craft could be controlled, descending upon its own exhaust jet? Drifting over the rock-filled Sea of Tranquillity, Neil Armstrong kept seeking for a place to land — having to control its descent manually, as the craft's computer system had switched off through overload! They made it with barely seconds left to go before the fuel would have run out. No wonder Houston radioed back, 'We're breathing again.' Earth was directly overhead.[51]

The chart shown for Houston (3.17 p.m., Houston time) has *no* squares or oppositions, as would signify impediment and stress, and is woven *entirely* from trines and sextiles, so no wonder it all went smoothly!

At the apex of the triangle is the Jupiter-Uranus conjunction, at zero degrees of Libra — the Sun's position at the autumn equinox,

Chapter 9. When Inventions Worked

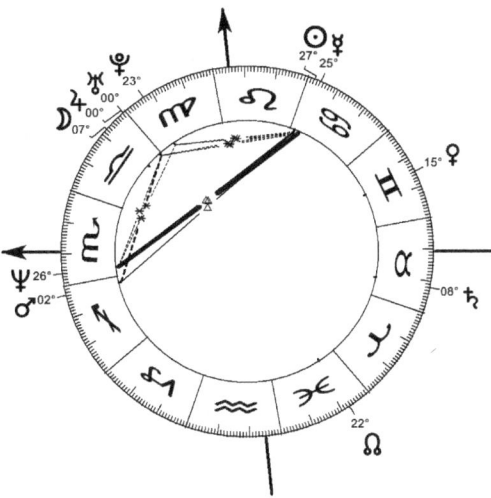

Figure 9.24: Event Chart for Lunar Landing, 20 Jul 1969 15:17hrs CDT Houston, Texas, USA. (5° orbs used.)

opposite zero Aries — midway between the Moon and Pluto, the nearest and furthest 'planets' of our solar system. I suggest this is *the* chart for the Dawn of the Space Age. Had anyone chosen this marvellous chart, one would admire their sagacity, but they didn't. It has Sun conjunct Mercury, for a travel adventure, and Sun trine Mars, for men using a fiery iron apparatus. Saturn was in biquintile to the Uranus-Jupiter conjunction. At Houston, Texas, Neptune conjunct Mars was rising, for the fulfilment of dreams through an iron apparatus. This beautiful chart helps to reassure us that the event did really happen (as it is doubted in certain quarters).

Gene Therapy
1990, 14 SEP ♅△♂, ♅△♀

A new era in medicine dawned at Bethesda hospital in Washington D.C. when a four-year old girl was treated by a transfer of genetic material. She was severely immune-deficient from a rare hereditary disease, due to a missing gene, leaving her a helpless prey to diseases. The new invention was a virus cleverly stripped of its harmful capability and having a copy of the missing gene. This gene had earlier been cloned and so it was quite well-known. The virus was capable of entering a human cell and donating this gene to the human DNA. The idea had to pass stringent medical, legal and ethical review boards, because of its far-reaching implications.

When permission had been granted, the paediatrician Ken Culver muttered, 'Well, here goes ...', and performed the injection, at 12.52 p.m., September 14th (16:52hrs GMT). It lasted half an hour. Over the following months, the girl's immune system slowly recovered, and she is now living a normal life.[52] There were five trine aspects present at this moment of success (to 5° of orb), which is, as we shall see, fairly relevant.

****

Soul of the Inventor

Leonardo da Vinci, Robert Hooke, Thomas Edison and Nicola Tesla — these are, by any standards, four of the greatest inventors who ever lived. By a happy chance, all have known birth data. Is there anything these charts share in common? There is, as it happens: a septile to Uranus. Leonardo da Vinci had Mercury septile Uranus, plus a Mercury opposition Saturn.

Robert Hooke, inventor of the microscope and pocket watch, and of much else besides (though he tended to lose credit for his inventions), had his Uranus in square to Venus, sextile to the Sun, septile to the Moon, and in decile (30°) to Mercury, all within 1°. He had plenty of Mercury aspects, appropriate for one buzzing around the place and communicating with all sorts of people and getting far more done in a day than one would normally think possible. His Mercury-Saturn trine was fine for a life dedicated to science. As a noon child his Sun was on the MC to 1° of orb, appropriate for one who lived his life on the public stage — but, he had no proper Jupiter aspects such as might tend to promote a decent public image, something he never managed to sustain, despite his great brilliance. Nor did he have any decent Mars aspects, such as might have given him the persistence and stamina required to actually finish and bring to fulfilment a few of the notions, which teemed so tumultuously from his fertile brain. All he had was Jupiter square to Neptune and Mars trine to Neptune, not a great help in these matters. The square to Uranus as his main

Chapter 9. When Inventions Worked

aspect to Venus points to a kink in his character whereby he entered into an incestuous relationship with his young niece Grace, and as he was living at Gresham College as Curator of the Royal Society this became a source of scandal. He is remembered in history books mainly for the enmity of Newton.

Edison had Mercury in septile to Uranus. 'Where do ideas come from? They must come from somewhere', he remarked. In his case they probably came from his Sun conjunct Neptune and Mercury, while his Jupiter in square to Saturn would have ensured they were of a fairly profound nature. That strong square in his chart may remind us of his remark about genius being 1% inspiration and 99% perspiration. He had a Mars-Saturn sextile (1°), appropriate for a scientific temperament, and indeed his remorseless driving energy which imposed strenuous demands upon his employees.

What Asimov called 'the climax of Edison's life', as 'nothing quite so dramatic ever happened again,' which brought reporters from all over the world to marvel, and which enabled the human race to lie a-bed for a few more hours every morning, was his lighting up of a filament electric lamp. This happened on October 23rd, 1879, when, as with Tesla's great electrical moment, strong Uranus-transits were stirring his natal chart. Saturn stood exactly over his natal Uranus, while Uranus was in opposition to his natal Saturn, and Jupiter was transiting his Venus-conjunct-Saturn.

The great electrical genius, Tesla, was born with Venus in such a septile to Uranus. His Mars was in biseptile to Saturn, and in biquintile to Uranus. Mars and Venus are the two planets we associate with the generation of electrical energy, as they have affinity with the metals iron and copper respectively, used in electrical apparatus. The dominant feature of Tesla's chart was its strong square between Mars and Venus: Venus conjunct the Sun, square to Mars conjunct the Moon.

There is a date associated with Tesla initiating the rhythm of alternating current: 16th May 1888, a day when Mars was in conjunction with Uranus. In New York on that day, he gave a lecture-demonstration showing his 'polyphase' electrical system, whereby dynamos could run on alternating current, something that had hitherto been viewed as impossible. Some powerful transits were then activating his natal chart: Mars-Uranus was over his natal Moon, and so in opposition to his Jupiter; while the Sun was conjunct his natal Uranus. He was having two different Uranus-transits on that day.

Glasgow's greatest inventor, John Logie Baird, who invented television, may not be in quite the same league as these other figures, but at least his birth time is known, as can be said of few great inventors. Did he have a septile to Uranus? He did have the Sun in such a septile, if we allow a wider orb than usual. It is conventional to allow a wider orb for solar aspects, and this septile is at 2° 12′. The mould-breaking quintiles were also strong: Sun-Moon and Mars-Venus, appropriate for an electrical genius. Quintiles are said to relate to one's art, to what one likes making.

References

1. James Burke *Connections* 1978, p.75.

2. Tom Tucker, *Bolt of Fate*, Benjamin Franklin and the electric Kite Hoax, NY 2003.

3. I. B. Cohen, R. Schofield, 'Did Divis Erect the First European Protective Lightning Rod?' *Isis*, 1952, pp.58–64.

4. Franklin Centenary Celebrations, 1906, Grand Lodge of Pennsylvania (copy in Royal Soc. library); I. Bernard Cohen, 'Benjamin Franklin's Two Lightning Experiments.' *Proc. Amer. Phil. Soc.*, (96) 1952, pp.331–336.

5. Larsen, *Ideas and Inventions* 1960, pp.172,178.

6. *Philip Reis, Inventor of the Telephone* by S. P. Thomson, 1883.

7. *The Talking Wire, the story of Alexander Graham Bell*, O. J. Stevenson, p.100.

8. R. Bruce, *Alexander Graham Bell and the Conquest of Solitude*, 1973, p.178.

9. Asimov, *BEST*, p.365.

10. *Reader's Digest Inventions that Changed the World*, p.288.

11. R. W. Clark, *Edison, the Man who Made the Future*, 1977, p.73.

12. *Thomas Edison, Chemist*, B. M. Vanderbilt, Washington, 1971, p.351.

13. *Quasar, quasar, burning bright*, 1978, N. J., Asimov, p.92.

14. *Edison*, G. S. Bryan, 1926 NY, p.115.

15. W. Dickinson, *Life and Inventions of Edison* 1895, p.195.

16. *Edison's Electric Light, Biography of an Invention* (New Jersey) by R. Friedel

and P. Israel.

17. R. Clarke, *The Scientific Breakthrough — the impact of modern invention* 1974, p.66.

18. J. E. Walsh, *One day at Kitty Hawk* NY 1975, p.138.

19. F. Howard, *'Wilbur and Orville'* NY 1978, p.138.

20. Walsh, op. cit. (18), p.272.

21. Beveridge, *Seeds of Discovery*.

22. *Mundane Astrology*, Baigent, Campion and Harvey, p.205.

23. (C. Harvey, 'The Galactic Centre', *The Astrological Journal*, Spring 1983, pp.74–84).

24. *The Discovery of Insulin*, Michael Bliss, p.120.

25. Larsen op. cit. (5), p.75.

26. *Rutherford, Recollections of the Cambridge Days* M. Oliphant, 1972.

27. *Lawrence and his Laboratory*, Heiltron and Seidel, 1989, Oxford.

28. R. W. Clarke, *The Greatest Power on Earth*, p.28.

29. *Cockroft and the Atom*, Guy Hartcup and T. E. Allibone.

30. C. Susskind, Who Invented Radar? *Endeavour*, 1985, 9, p.93.

31. *Howard Florey*, TIC. Williams, 1984, p.81.

32. J. Golley, *Genesis of the Jet, Frank Whittle and the Invention of the Jet Engine*, 1996.

33. Richard Rhodes, *The Making of the Atomic Bomb*, 1986, p.442.

34. E. Troinsky, *Das Horoskop des Atom-Zietelters*, 1975, Munich, pp.92–3.

35. Mark Lerner, 'Atomic Energy and Astrology' *Welcome to Planet Earth*, 1985 (Jnl) Summer (Oregon, US). Lerner first gave the chain-reaction chart.

36. W. Shockley, The Path to the Conception of the Junction Transistor, *IEEE Transactions on Electron Devices*, 1984, 11, pp.1534–7.

37. R. Lee, *The Eureka! Moment, 100 Key Scientific Discoveries of the 20th Century*, The British Library 2002, pp.236–7.

38. *Alan Turing, The Enigma of Intelligence*, Andrew Hodges, p.385.

39. Prof Kilburn at Manchester gave the information about time of day for this event.

40. S. Movington, *History of Modern Computing*.

41. *Achievement in Radio*, Snyder & Bragaw, National Bureau of Standards,

Boulder, Colorado, 1986, p.293.

42. Robert Jungk, *Brighter than a Thousand Suns*, p.262.

43. Rhodes, op. cit. (32), p.480.

44. *Bulletin of the Atomic Scientists*, 'The Hydrogen Bomb Story', December 1952.

45. E. Troinski, op. cit. (33) assumed an 11hr timezone, p.102. I am here indebted to Helen Clerf of Wyoming for her sleuthing to locate this time, as also several others.

46. Koestler, *Janus: The Summing Up*, 1978, Ch.1.

47. *Lasers & Applications*, May 1985, T. Maiman interview, 'Laser Pioneer' pp.85–90; I. Flatow, 1951, pp.122–4.

48. Koch, *The Creative Engineer*, 1978, p.260.

49. Emmett N. Leith, Some Highlights in the History of Display Holography, U. of Michigan, Ann Arbour MI (undated); press release from *Amer. Inst. Phys.*, 5.12.1963.

50. R. M. Hazen, *Superconductors: The Breakthrough*, 1988, p.52.

51. Andrew Chaikin, *A Man on the Moon, The Voyages of the Apollo Astronauts*, 1994, p.200; six hours later (9.56 p.m. Houston time), they set foot on the Moon.

52. N. Lemoine & D. Cooper, *Gene Therapy*, 1996, p.5.

CHAPTER 10

Structure of the Trine

'The number Three is related to the idea of life, vitality and enjoyment, and hence to what motivates us and moves us to action.'[1]

Mundane Astrology, Baigent, Campion and Harvey.

WHAT FORM HAS A CELESTIAL ASPECT? We seek for the distinctive pattern formed in the heavens, as new inventions came into being. We began our investigation of this matter in chapter 7, and now return to it using a larger data-set. Will these support the earlier findings?

As before, results are depicted graphically. The method compares the *astronomically* expected frequency in a given setup with the empirically found frequency. Any difference between the two may have an *astrological* cause. The machinery of the solar system generates a certain frequency of events, over a given period, which the computer models. If this is substantially lower than what is found, we may wish to say that the difference is 'significant' — i.e. that some kind of *influence* is at work, whereby the timing of events has deviated from chance.

The computer does all the number-crunching. One may compare the mathematics in these analyses to what is under a car bonnet: we need to be confident that it is running properly, but otherwise would prefer not to know the details! If some keen readers wish to re-check these analyses, this is now becoming feasible using the 'research' astrology programs on the market.*

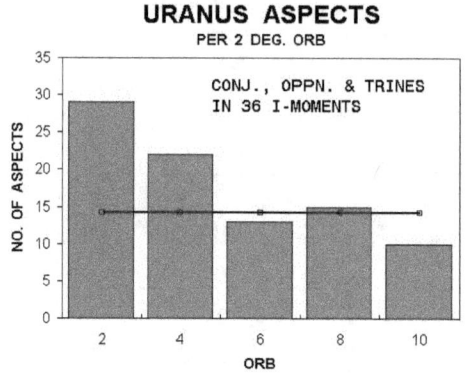

Figure 10.1: Major aspects to Uranus within the Invention-moment group scored per 2° of orb, and compared to chance-expected level. There is an excess of 100% within the first two degrees of orb.

In chapter 8: Uranus the Awakener, a massive sixty percent excess of major Uranus aspects appeared, in the I-moment group. We found this using the major aspects i.e. conjunction, opposition and trine, set to five degrees of orb. Now, we are going to check out how these Uranus aspects varied with orb: how close were they, and what would have been the best orb we could have used?

We start by grouping all of these major Uranus aspects together, and plotting them at 2° intervals of orb, to give the distribution shown, where the largest excess appears as within two degrees of orb, and all of the excess as within four degrees. This provides further evidence for the effect being a valid one. Were it an artifact or statistical quirk, then one would not expect an enhancement of the effect at diminishing orb. Four degrees appears as being the optimal orb.

This implies that the analysis conducted in chapter 8, would have been better done using just four degrees of orb. Five degrees was chosen fairly arbitrarily, as it had been used by others (e.g., the

---

*The program Jigsaw by Astrolabe will output chance-expected frequencies over a specified period, for given conditions. They clock up random charts from which 'chance-expected' values are derived. One really needs astronomical advice about such expected frequencies, which can be tricky.[2] Mars, for example, spends longer conjunct the Sun than it does in opposition, making its pattern in the diurnal cycle quite a headache to grapple with.[3] A 'solar clustering' effect causes the inner planets to have slightly less trine aspects *in general* than the outer planets (appendix B). The best-known 'research' program, *Jigsaw* by Esoteric Technologies Pty Ltd/Bernadette Brady, can sum traditional (Ptolemaic) aspects within groups of data, and also the quintiles and septiles.

Gauquelins) in investigating these major aspects. Earlier, in the case of the quintile and septile aspects for the natal eureka group, we found that the optimal orb was smaller than that chosen for the analysis. In both cases we would have been better off using a tighter orb, which is encouraging. The aspects 'chime' over a quite limited angular distance.

Harmonic Analysis

A harmonic analysis, as we saw earlier, scores all aspects grouped within a given 'harmonic'. Thus, a 'sixth-harmonic' defines angular relations that are multiples of 60° between two points, excluding the zero position. There are five of these, which come from adding up the trine, opposition and sextile. The computer performed such an analysis up to the fifteenth harmonic.

As before, 'Addey orbs' were used having a base-orb of twelve degrees, so that the orbs for each harmonic were $(12° \div n)$ where n is the harmonic number. Thus the second harmonic (opposition) will be six degrees, the fourth harmonic (squares) will be three degrees and so forth. This formula gives us quite reasonable orbs except for the conjunction: an astrologer would never use twelve degrees for a conjunction while taking six for an opposition. The last column gives chi-squared values, indicating roughly the significance of values.

This table *discovers the trine excess*. Until this analysis, one had no idea that trines were by far the strongest harmonic in the I-moment group. The chi-square value for this excess is 14, as would have been significant at several thousand to one, had it been predicted in advance. (For comparison, chi-squared values above four are statistically significant). They are showing nearly a fifty percent excess! After such a strong effect, it is not surprising that the sixth and (to a lesser extent) the ninth harmonics are also raised, being multiples, or 'overtones', of that third harmonic effect — the sixth harmonic includes the two trine positions at half the orb.

The other two harmonics significantly in excess are the 7th (septile) and the 14th. This result is comparable to that found in the E-moment group, where higher 'overtones' of the seventh harmonic were very much present. One would like to see a more accurate computer re-analysis of these high harmonics in the two data sets,

Table 10.1: Celestial aspect frequencies during the 36 I-moments.

Harm.	Aspect	Score	Expected	Difference	Chi-Square
3	trines	92	62	+30	14 (48% excess)
4	squares	72	72	0	0
5	quintiles	64	79	-15	3 (19% deficit)
6	sextiles	102	84.5	+18	4 (21% excess)
7	septiles	107	88	+19	4 (22% excess)
8	octiles	91	90	+1	0
9	noviles	97	91	+6	0
10	deciles	93	92	+1	0
11	undeciles	97	93	+4	0
12	dodeciles	91	94	-3	0
13		74	94.5	-21	4 (22% deficit)
14		117	95	+22	5 (23% excess)
15		96	95	-1	0

because of the very small orbs of the 'overtones'. If confirmed, this could endorse some of Addey's rather speculative comments about these fine, high-number harmonics.

The fifth and thirteenth harmonics are in deficit. The quintile aspects, so important for the eureka moments, would appear to be detrimental for the more practical moments of invention. The deficit in the thirteenth harmonic may confirm the traditional view of thirteen as an unlucky number! It appears that the prime numbers 3 and 7 tend to be present at the creation of new forms in science or technology.

What is the optimal orb for a trine? Answers given hitherto to this millennia-old question have been little more than arm-waving conjecture. Firstly, the computer summed the trine aspects over degrees of orb, and found the results as shown in figure 10.2 (a). This 'aspect profile' graph shows that the 'Addey orb' of 4° well fits the trines. Next, the remarkable 'time-structure' graph of figure 10.2 (c) separates applying and separating trine aspects, demonstrating the startling fact that the trines exert their main effect *before* exactitude. These resemble the pattern found for the septiles: figures 10.2 (b) & (d). Figure 10.2 (c) also shows a powerful *deficit* in aspects *prior* to coming within orb, comparable in amplitude to the *excess* of aspects *within* orb — indicating that the inventors had to wait until the trines were in place for their inventions to work!

Chapter 10. Structure of the Trine

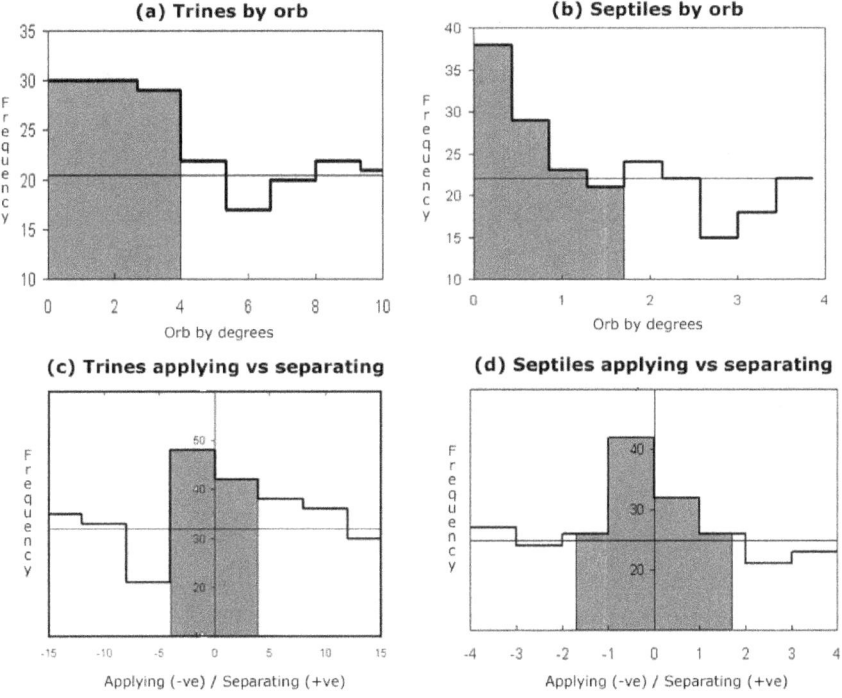

Figure 10.2: Trines and septiles in the group of 36 I-moments.

Return of the Septiles

There was another big surprise: again the septiles turned up, showing virtually identical characteristics to those in the eureka group, though somewhat weaker. If the E-moments express the septile quality of inspiration, the I-moments were events of action, with their energy expressed by the trine motif. The septile graphs show the same prevalence at close orb as did the E-moments, which is evidence for the validity of the effect we are looking at: increasing strength with closeness of orb would not be expected from an artifact.

Figure 10.2 (b) shows the septile excess within the I-moment group. Nearly all of the excess was within the first half of the 'Addey orb', or just under one degree ($\frac{6}{7}°$), showing a 50% excess in the score of I-group septiles. This is a closely similar result as was found earlier for the 21 E-moments, where a 90% excess of septiles appeared at that close orb — a somewhat stronger effect. The E-moment group

225

showed a pure expression of septile-inspiration, having little of any other harmonics in excess, while the I-group had two major harmonics, trine and septile. In each group the septiles showed a larger effect at close orb.

The 'time-structure' graph for the septiles in figure 10.2 (d) showed the same excess prior to exactitude that we saw in all the other such graphs. This is the fourth time such a pattern has appeared. Science progresses by discovering similar pattern in different realms. The notion that an aspect has more energy or tension prior to exactitude than after, has been common amongst astrologers at least in recent centuries.

Again, the 14th harmonic has turned up strongly in the invention-group — in fact it had a larger amplitude than the 7th. As harmograms come to be used to plan future occasions of inspiration, this probably means that an input for the 14th harmonic would feature, as well as the 7th. To get a feeling for the 14th harmonic it may help to turn back to the heptagon family as shown in chapter 4: The Angles Of Inspiration (figure 4.6). The glorious star-heptagon makes the one-fourteenth angle, while angles at the corners of the other heptagon figures are multiples of this one-14th of a circle, i.e. just over 25°. These form what John Addey would have called the 14th-harmonic *family or series*.

As the discerning reader will note, we are here at the limit of the smallest circle divisions that were permitted as effective aspects by Kepler, as quoted earlier: those aspects 'which hesitate between power and powerlessness.'[4] One may experience the 14th harmonic as resembling 'glittering irradiations', surrounding the sevenfold pattern.

References

1. Michael Baigent, Nicholas Campion and Charles Harvey, *Mundane Astrology* 1984, 1992, p.149.

2. For discussion, see 'Investigating Aspects' by N.K. in Pottenger Ed., *Astrological Research Methods*, CA 1995, pp.287–302.

3. Pottenger, 'Mars Days', *Kosmos* Winter 1997 pp.36–45, (though diurnal frequencies are not relevant to the present study).

4. Footnote p.22, p.52; Kepler, *Harmonice Mundi* 1618, 1997, p.347.

Part III

Perspectives

CHAPTER 11

Destiny and the Uranus Cycle

'Those who do not know the torment of the unknown cannot have the joy of discovery'

Claude Bernard

The Transits of Uranus

URANUS TAKES EIGHTY-FOUR YEARS to orbit around the Sun. This period measured the life of Sir William Herschel, who discovered the planet. He died on his Uranus return within one degree, and also with Neptune opposition Neptune (0° 15′), reminding us that Neptune's orbital period (164yrs) is almost twice that of Uranus. Herschel discovered the new planet Uranus just halfway through his life: born on 15th November 1738, he made his discovery in March of 1781. That corresponds to the 'opposition' stage of the Uranus cycle, of 180° between the natal position of a planet and its position in the heavens.

If the discoverer of the new planet so well exemplified its cycle, then perhaps we should take notice of its occurrence in the lives of exceptional scientists. William Herschel was a musician, interested in music theory and Pythagorean philosophy, so perhaps he would have appreciated the following: the number 84 of its period is the product of 12 and 7, so that the major life-stages for which these aspects chime are multiples of seven years — 21 years for the Uranus square aspect, 28 years for the trine, and 42 years for the opposition.

Taking an orb of 3° for these key positions gives a period of a year and a half per transit.

The planet continued to be called Herschel all throughout the nineteenth-century, as its nature came into focus within the minds of astrologers. As astronomers ascertained its orbit and motion, astrologers were determining something more important, namely its character and being. For this they used not merely the rational mind, but their intuition.

We may here reflect upon initial impressions of that sphere's influence by astrologers. At first it was called 'Herschel.' An early view was that of 'Zadkiel', who wrote that:

> The nature of Herschel is extremely evil ... of very eccentric disposition ... one who despises the track of custom.[1]

Later, W. Simmonites found that 'Herschel' was 'Full of inventions and novelties,' while in 1898 one author found it to be:

> Active in manner, fond of study, also fond of science, inventive and fond of the occult or mysterious ... When well dignified, he will probably make some invention or discovery that will cause him to become famous or very wealthy.[2]

Dr Richard Tarnas, the American philosopher, has argued that the real nature of the planet Uranus is 'Promethean.' The figure of Prometheus was, he explained, 'the supreme mythic personification of rebellion, revolution, technological and cultural innovation, and the striving for freedom and change.'[3]

Tarnas has described how he first apprehended the significance of its transits while studying the life of Sigmund Freud. At the halfway point of his Uranus cycle Freud reached his deep insights, formulating his theory of dreams and the unconscious. 'Insight such as this falls to one's lot but once in a lifetime,' Freud later wrote.[4] His notion of psychoanalysis emerged in a paper completed in February 1896. A colleague of Freud's, Breuer, wrote of him in the summer of 1895: 'Freud's intellect is soaring at its highest. I gaze after him as a hen at a hawk.' Freud's biographer wrote that at this period, 'Freud was at his most revolutionary stage, both intellectually and emotionally.'

Chapter 11. Destiny and the Uranus Cycle

'This wave of events which coincided with Freud's Uranus opposite Uranus transit', Tarnas concluded, 'seemed to reflect a distinct climax in the unfolding of Freud's Promethean impulse, giving the impression that the opposition represented a kind of "full Moon" moment for the Prometheus archetype in the lifelong Uranus cycle. The correlations were sufficiently striking that I began to investigate cases of other major scientific revolutionaries, in each instance comparing the astronomical situation with the biographical data. In the case of Galileo, the classic Promethean figure in the history of science, I found that Uranus had reached the opposition point of its cycle during the period from July 1607 to June 1610 [N.B. this rather wide period corresponds to 5° orb]. It was in the fall of 1609 that Galileo had first turned the telescope to the heavens, there discovering the craters on the Moon, the moons of Jupiter, the numerous individual stars of the Milky Way, and other celestial phenomena supportive of the Copernican hypothesis, and it was on March 12, 1610 that he published Sidereus Nuncius (The Starry Messenger), the epoch-making account of his observations. During these several months Uranus was within 3° of exact opposition to its placement at Galileo's birth, and on the day of the book's publication Uranus was within 1°. Again it seemed that in the life of a Promethean individual, the midpoint of the Uranus cycle represented an unmistakably climactic moment in the unfolding of that individual's Promethean impulse.'[5]

To this luminously original thesis we may add, that when Galileo first looked at the moons of Jupiter, on January 7th, 1610, Uranus was closely conjunct Jupiter (2°), such that indeed one marvels that he failed to detect it with his telescope. Uranus was then opposite to its natal position within 2°. On another occasion he spotted Neptune as it passed by Jupiter, and was rather puzzled by it. Earlier, he first looked at the Moon with his telescope early in the evening of November 30th, 1609, when his Uranus opposition was within three degrees, and the Sun was conjunct his Uranus within 16 arc minutes.

In the *Starry Messenger* published in March of 1610 (with Jupiter still conjunct Uranus), Galileo told his readers how 'it is a most beautiful and delightful sight to behold the body of the Moon. The book was an instant bestseller and effectively introduced the telescope to Europe. He, Galileo, had first applied it to the skies. Nothing he ever published later had quite the same impact. Written in a terse, easily readable style, his *Starry Messenger* brought home the meaning of the new, sun-centred universe as Copernicus' long and unreadable tome never did:

> By the aid of a telescope anyone may behold this in a manner which so distinctly appeals to the senses that all the disputes which have tormented philosophers through so many ages are exploded at once by the irrefrangible evidence of our eyes, and we are freed from wordy disputes upon this subject, for the Galaxy is nothing else than a mass of innumerable stars planted together in clusters.

But also, his *Starry Messenger* described (in its Introduction) the chart of Cosimo II de Medici, as having Jupiter in a dominant position (at MC). Galileo proposed to call the new moons of Jupiter he had found the *Stellae Medici*, for this reason. From being a *mathematicus* at Padua university, he thereby became court philosopher to the Medicis in Florence, with them promoting his telescope! This grand apotheosis of his career was empowered by his Uranus opposition *plus* Uranus' conjunction with Jupiter. Later in that year, still within that opposition, he detected sunspots and the phases of Venus.

Tarnas looked at the philosophers of science, Descartes and Bacon, discerning that their major works laying the foundations of their new philosophy were published within one degree of their Uranus opposition returns: Descartes' *Discourse on Method* was published in June 1637, and Bacon's first major philosophical work, *The Advancement of Learning*, was published in October 1605, setting forth Bacon's vision as to how scientific knowledge was the key to human progress.

Chapter 11. Destiny and the Uranus Cycle

Newton's Uranus Cycle

The life of Isaac Newton is the textbook case-study for the Uranus cycle in the life of a scientist.[6] Living to what was then the advanced age of 84 years, he died with Uranus at $2\frac{1}{2}°$ from its natal position. His main creative periods fit with precision into the key phases of the Uranus/Prometheus cycle. We now trace out the major steps of his Uranus cycle.

Unlike most other scientists, his creative work began at the quarter-cycle, at the youthful age of 21. The first Uranus quarter-period normally signifies youthful rebellion against convention and tradition. In the case of Newton, a lot of his deepest scientific insights derive from this period.

Figure 11.1: Newton's prism experiment (*Mary Evans Picture Library*)

His biographers speak of Newton's 'Annus Mirabilis' as having been around 1665 and 1666. This was partly a literary device, because much was expected of the year 1666 and poets spoke expectantly of it as an 'annus mirabilis.' In the event there came the Great Fire of London and the Plague. Newton, born in December of 1642, was then a graduate student. His modern biographer Westfall referred to 'the years 1664 to 1666, [Newton's] *anni mirabiles*', adding that 'The autumn of 1665 passed with incandescent intensity'.[7] That defines the precise period of Newton's Uranus square. His student notebooks show the great struggles which went on, as infinitesimal methods in mathematics opened up for him, and the beginnings of his calculus method were formulated. He purchased a prism and then started his inquiries into light and colour. In later life, Newton made the much-quoted remark about this early period: 'I was in the prime of my age for invention and minded Mathematics and Philosophy more than at any time since.' As the Plague descended onto the university town, Newton left for his mother's house at Woolsthorpe, near Grantham, in August of that year, and did not return until March of 1666. The traditional tale of the apple was located in this period.

235

In the next decade, Newton's innovative theory of light and colour emerged. This was despatched to the Royal Society on February 6th, 1672, where it was read out. It was an immediate success, the Royal Society's Secretary Oldenburg informing him that his paper 'mett both with a singular attention and an uncommon applause.' This was a dramatic follow-up to the new telescope which Newton had presented them with, a few months before. These two events in close succession established his name before the cognoscenti of Europe.

His transits for that day were:

Uranus	trine	Uranus	2°
Saturn	conjunct	Saturn	0° 5'
Jupiter	opposition	Jupiter	1°
Moon	square	Moon	4°

Newton had a loose Saturn-Jupiter conjunction in his natal chart, in trine to his Uranus. Uranus had moved round to be on his Jupiter (1°), when it was conjunct Saturn and the North Node, and opposite to Jupiter, making an awesomely powerful transit over Newton's Jupiter-Saturn conjunction. Having the North Node aligned with the two Saturns is most appropriate for this event concerning the very foundation of modern science. (For an experiment which made a colour-image out of sunlight, it's nice to see that transits of Sun conjunct Venus *and* Venus conjunct Sun were then happening.)

Newton stepped onto the stage of history with this missive, prior to which he was merely a rather high-powered mathematics lecturer, and amateur chemist/alchemist. He was at once elected to become a Fellow of the new Royal Society.*

We are not suggesting that Newton had an astrological advisor. Rather, we are suggesting that fate, or perhaps Fate, caused him to take advantage of so highly appropriate a moment in his life's unfolding. Before sending this letter, when Oldenburg had written to him congratulating him on the reflecting telescope, Newton had replied that he was now ready to forward a 'Philosophical discovery' about light, adding that it was 'in my Judgement the oddest if not the most considerable detection which has hitherto been made in the

---

*Readers wishing to verify these charts will need to add on ten days to the dates given, to allow for Britain's remaining in the old Julian Time in this period, or make sure that their software will make the appropriate adjustment.

Chapter 11. Destiny and the Uranus Cycle

operations of Nature.' If that was an exaggeration, it was only a slight one.

His Uranus-opposition return brings us to the composition of his immortal *Principia*, linking earth and sky with the same laws of mechanics. That opposition became exact in the spring of 1687, when Newton was completing the grandiose Book III of the *Principia*. His colleagues saw him then as forgetting to eat his meals, and as silently drawing diagrams on the ground in the garden at Trinity College with a stick. He grappled with the Moon, gravity and comets. With tiger-like vigour, his mathematics grasped ahold of their motion.

Edmond Halley was midwife to the *Principia*, and carried the publishing costs himself. His Sun was closely conjunct Newton's Uranus, within 40′, and his Uranus was conjunct Newton's Sun (3° orb). He gave Newton the stimulus for composing the masterwork, and egged him on during it, checking over the proofs, announcing it to the world, and paying for the printing costs. Its production is viewed as Halley's greatest achievement! Its reception was formally announced in the Royal Society's journal-book as for April 28th, 1686 — that date being Halley's own Uranus-trine return to 17′. As it happened, this was merely for the first two books of the *Principia*, for Book III had still to be written.

The *Principia's* announcement at the Royal Society saw a Full Moon aligned with a Uranus-Jupiter opposition, all focused on Newton's natal Uranus position, conjunct Halley's Sun! No wonder it was Halley's big moment.

Halley had to placate Newton when, on the first announcement of the opus to the Royal Society, Hooke claimed that his ideas had been used without proper acknowledgement, causing Newton to storm furiously through much of that summer. Having Jupiter passing over his Sun may have helped Halley at this moment.

In March of 1687, as Halley's Saturn return chimed, a letter from him to Newton concluded, 'I am much rejoyced to be any wais concerned in handing to the world that that all future ages will admire.' Uranus had by then come into conjunction with the lunar nodes, as it remained in opposition to his Sun and Newton's Uranus, within a degree. The magnum opus was completed. Tarnas' conception of the Uranus transits gives us a way of comprehending the unfolding of Newton's genius. The half-orbit position occurred in May of 1687,

237

and the *Principia* was published on 5th July. As Newton was born in December of 1642, this was a passage of 44 years and five months, rather more than half of the complete period of Uranus of 84 years. Uranus does not move at uniform speed, having quite an elliptical orbit.

The opposition return, defining the midpoint of his life, marks a huge division in his life which is quite hard for his biographers to cope with. Prior to it, Newton was a most reclusive scholar with an ivory-tower job in academe, immersed in private alchemical experiments which no-one has ever understood, entirely removed from ordinary human affairs. Almost the only anecdotes about him concern his absent-mindedness. Virtually all of his creative scientific work was achieved in this period. Then, suddenly, he transforms into a public figure. Hardly had he completed the *Principia* in March of 1687, when Newton urged the Fellows of his college to resist the imposition of a Catholic representative as the Monarch had ordered. This act of resistance put his career at risk, but paid off as King James fled the country two years later. In consequence of this bold act, Newton was asked to represent Cambridge in the new Parliament of 1688. The rest of his life saw him politically involved in one way or another, becoming President of the Royal Society, Master of the Mint and then knighted.†

Britain's best-known scientist chimed in well with his next Uranus transit. He had forsaken the groves of academe for the grime and bustle of London, taking a job at the Royal Mint, when his second trine transit arrived in August of 1701. He was then Master of the Mint, answerable to the King of England for the state of the nation's currency. Newton was promoted to his position of the Master of the Mint in 1700 when the Bank of England had just been established and the first paper money was being floated. A contract was struck between the King and the Master of the Mint in December of 1700, which 'embodied the supremacy of the master over the Mint' ... and 'specified the terms under which Newton as Master would mint — for example, the coins he would issue, their weight etc.'[8] At this grave moment in his life, he was supported by his Uranus-trine return at 2° orb.

---

†His nervous breakdown of May–September 1693 happened as Neptune transited over his Saturn.

Chapter 11. Destiny and the Uranus Cycle

The year 1701 was a busy one for the Mint, and Newton's salary for that year amounted to £3,500. He finally resigned his fellowship at Trinity and also his chair as mathematics lecturer, as he had stopped giving lectures there years ago. He had to face a public trial of the accuracy of his coinage, called 'the Trial of the Pyx'. In theory he could be sued if there was not enough gold in the coins. Various lords representing the King and governors of the Goldsmiths Company assembled in the Star Chamber, on the morning of August 6th, 1701, and Newton stood before them. The exact quantities of gold and silver in sample coins were tested, and Newton's currency passed the test. His trine transit again stood at 2° orb. Afterwards, there was a large feast at the Dog Tavern to celebrate. While this has little relevance to the theme of scientific discovery, it shows how this particular cycle continued to work in his life, defining its climactic moments.

Moving on to the second square aspect in late 1707, this saw Newton at 65 returning to natural philosophy, again serving at the Mint and as a Member of Parliament, becoming President of the Royal Society and knighted by Queen Anne. He then started preparing a new edition of his immortal *Principia* — doubtless not so striking an act as in his previous Uranus/Prometheus transits, yet in accord with them. Thus we see how, at the first quarter, the first germs of his philosophy appeared, at the opposition, he composed his magnum opus, then at the second quarter he began revising it, adding in some new theories.

At the full cycle return, in the year before his death, he gave form to a myth. In 1726, William Stukeley came to visit Newton at his house in Kensington, whence he had retired to escape the fumes of London. Twenty years later on, it was to dawn upon Stukeley that Stonehenge was aligned with the midsummer sunrise. He much admired Newton and composed the first British biography. He found Newton hard at work on his *Chronology of the Ancient Kingdoms, Amended.* They had tea in the garden.

It was in such a garden as this, recollected the ageing sage, that the notion that gravity might extend as far as the Moon first came to him; why, it was even as an apple dropped from the tree, that the notion formed. That was the first documented version of the apple myth, told to William Stukeley! In biographies which emerged after Newton's death, the next year it was related, and soon caught on. Earlier versions of how the notion of gravity theory came to him were about, dating from shortly after his nervous breakdown of 1693, however no apple was present in them. The immortal story of the falling apple and a garden first emerged upon Sir Isaac's Uranus return! Thus was the best known of eureka stories created.

We have looked at the lifelong friendship between Edmond Halley and Isaac Newton, the former being deeply involved in working out Newton's destiny. We see this expressed in the alignment of major oppositions in their two natal charts, that are together *within a single degree*: Halley had Jupiter opposite Venus at 3° Gemini and 1° Sagittarius, reflecting his gregarious and social disposition, while Newton had Pluto at 2° Gemini and Neptune at 1° Sagittarius, an opposition that only happens once in centuries. Briefly, let's take a look at Halley when his own Uranus opposition chimed: he was captain of a ship exploring Antarctica, amongst penguins and deadly icebergs. The Admiralty had commissioned him to explore the *Terra Incognita* and he went further South than anyone had done before. When Halley returned from this epic voyage, his colleague Flamsteed complained that he 'talks, swears and drinks brandy like a sea-captain.' However, that did not prevent him from being appointed geometry professor at Oxford (with Newton putting in a word for him). Thus, his Uranus opposition was one of grand excitement and adventure. We find a pattern in their destinies by using a planet then unknown. These pioneers of a new science sought for an 'objective' theory, and kept clear of self-knowledge or an inner understanding of things. If we can now supply that, it is only by reference to a then-invisible sphere.

Ben Franklin

> '*And stoic Franklin's energetic shade
> Robed in the lightnings which his hand allay'd.*'
>
> Byron, *The Age of Bronze*

Benjamin Franklin 'snatched the lightning from the heavens and the sceptre from tyrants', wrote the French economist Turgot. Dr Tarnas has given us a language for describing the development and crucial experiences of those heroic, larger-than-life figures of destiny whom he calls 'Promethean', a thing which modern psychology greatly lacks.

The book, *Franklin and Newton* by science historian Bernard Cohen argued that there were various points of comparison between the lives of these two towering geniuses. Franklin was the only one of the founding fathers of the American nation to achieve a European reputation. He had a mercurial versatility, being in turns a printer, writer, politician, diplomat and scientist — linked by Tarnas to Franklin's (rather wide) Mercury-Uranus opposition. He lived as though he always knew what the next step of his destiny was going to be. The eighty-four years of his life spanned an exact Uranus cycle.‡ His death-transits were:

> Uranus conjunct Uranus: 2° 10′
> Pluto opposition Pluto: 0° 38′

One is startled to find a Pluto opposition Pluto at his death, as Pluto's orbit period is thrice that of Uranus, not twice. Its highly eccentric orbit comes within that of Neptune for a while. It remained there over the middle period of Franklin's life, reaching perihelion§ in 1741, a few years before Franklin immersed himself in the theory of electricity. In May 1752, as his name became celebrated throughout Europe from his lightning conductor experiment, Franklin's natal Uranus was receiving an exact trine from Pluto.[9]

Franklin's interest in electricity was awakened in 1746, chiming exactly with his Uranus-opposition transit over 1746/7. These were the two years in which his theory of electricity developed. Of the year 1747, his biographer said: 'Propounded his theory of electricity known as the Franklinian Theory', and then for 1748: 'Propounded his theory of the electrical condition in the Leyden jar.' In March of 1747, Franklin wrote to his friend Peter Collinson in London that:

> I never was before engaged in any study that so totally engrossed my attention and my time as this has lately done,

---

‡Franklin's birth date, January 17th, 1706, is New Style. America as a whole adopted the new Calendar system in 1752 as did Britain, but Boston switched over somewhat earlier.

§Perihelion is a planet's nearest approach to the Sun. At the opposite end of the orbit is aphelion, when it is furthest away.

> for what with making experiments when I can be alone, and repeating them to my Friends and Acquaintances, who, from the novelty of the thing, come continually in crowds to see them, I have, during some months past, had little leisure for anything else.

Collinson passed on Franklin's letters to the Royal Society, and it was through these that his work became first known in England. Franklin's Uranus-opposition return was then beginning at $4\frac{1}{2}°$ of orb, as he started performing his experiments with a Leyden jar. On the eleventh of July, as Uranus was transiting his natal Mercury ($1°6'$), Franklin wrote a new letter to Collinson expounding his historic new theory of electricity: how the different types of electricity were like a single fluid, how the terms 'positive' and 'negative' should be applied, and the 'wonderful effect of pointed bodies.'

In September, when Uranus was a mere three degrees away from its opposition, another letter to Collinson unfolded his theory of the Leyden jar. His electrical writings were published in 1751. They caused a great stir, and were admired for their lucidity. 'He had endeavoured to remove all mystery and obscurity from the subject', was Sir Humphrey Davy's comment.

Both Newton and Franklin had inspiration dawn at the midpoint of their Uranus-orbit lifespans, right over the opposition. A reviewer in the *New York Review of Books*, discussing a spate of new books on Franklin's life and work, commented on how he:

> retired from business at the age of 42, prepared from then on to spend what he had got rather than to accumulate more. For his remaining forty-two years — he happened to divide his lifetime neatly in half — he had no need to think about making or saving money.

Relativity Dawns on Einstein

The twentieth century was ushered in by a slow, long-lasting opposition between the two outermost known spheres, Uranus and Neptune, lasting from 1905 to 1910. This gave birth to the new Theory of Relativity, together with cubism in art, atonal music and free-association

Chapter 11. Destiny and the Uranus Cycle

in literature. Few could understand the new theory, and the new art-forms were quite a strain.

> The year 1905 was the *annus mirabilis* both for Einstein and for physics. It was in that year that Einstein, at the age of twenty-six, published three papers, each epoch-making in its own way.[10]

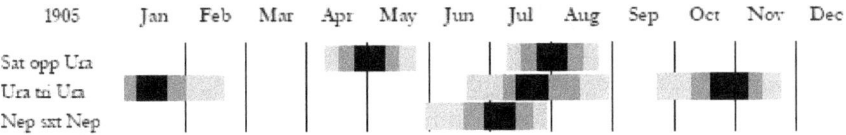

Figure 11.2: Einstein's 1905 transits.

When asked in later years what was special about his mode of approaching physical problems, Einstein would stress his ability to indulge in nonverbal thinking. As a youth, he used to dream about travelling on a beam of light and wondered how the world would appear from that vantage-point. One physicist said that, 'In a kind of miraculous way, in general relativity, the curvature tensor in Reimannian geometry and the energy momentum tensor in physics joined.' In this treatise, we need not describe what it was that came to the great scientist! All the evidence anyone needed to frame the new theory had been around for at least fifteen years, but it could not be envisaged until these two spheres came into their opposition.

In that year 1905, Albert Einstein was walking in the Swiss Alps, near Berne. He was paying a somewhat unexpected visit to his friend Besso. They had a productive discussion, and many years later Einstein recalled how:

> I could suddenly comprehend the matter. Next day I visited him again and said to him without greeting, "Thank you, I've completely solved the problem." My solution was really for the very concept of time ... With this conception, the foregoing extraordinary difficulty could be thoroughly solved. Five weeks after my recognition of this, the present theory of special relativity was completed.'[11]

¶Of this day, Einstein later recalled: 'That was a very beautiful day when

If that laconic remark was all Besso was told, it is hard to blame him for not recording the epochal date in his diary. Somewhere around May 20th, 1905,[13] Time and Space were fundamentally altered, indeed were never the same again. Einstein's completed paper on what later came to be called the Special Theory of Relativity was received on June 30th.

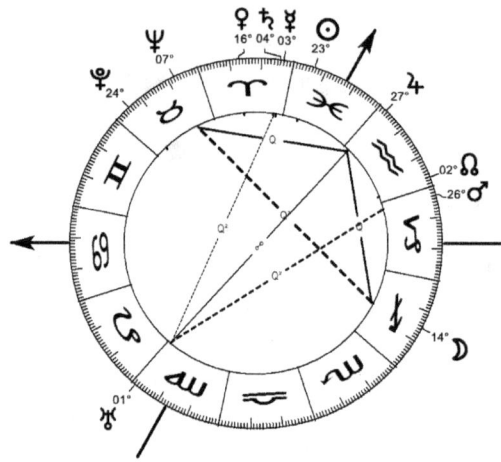

Figure 11.3: Natal Chart of Albert Einstein, showing quintiles plus opposition aspects, 14 Mar 1879 NS 11:30hrs LMT Ulm, Germany. (5° orb opposition/3° orb quintiles used.)

Uranus the innovator was pulling in opposition to Neptune. For a theory which dissolved the intelligibility of the universe, Neptune-Uranus has to be the appropriate combination. Before the strange birth at this opposition, Matter was Matter, Space was Space and Time was Time, but after it they had all somehow changed. Potent new equations, incomprehensible to all but a select priesthood of the

---

I visited him (Besso) ...' Pais remarked that Einstein's total concentration on relativity followed immediately upon his completing several other projects, the last of them on May 10th.

[In 1922, Einstein gave a lecture in Japan, when he *first* related the above incident. One expert[12] has advised the author that, 'unfortunately, I never realised any diaries of anyone which can fix the day of the meeting between M. Besso and A. Einstein in Bern,' and cautioned that there was no definite evidence for the surmised date of May 20th. But, assuming that he posted off the paper upon completing it, then the weekend of 20th–21st May seems a reasonable candidate for his eureka insight.

Chapter 11. Destiny and the Uranus Cycle

new arcana, warped the fabric of the old Newtonian universe. By the completion of the theory on June 30th, that opposition had moved (temporarily) out of aspect, i.e. beyond 5°, as the birth process was completed. This was the first such opposition since the discovery of Neptune the previous century.

'...in his student days, Einstein had been a lazy dog. He never bothered about mathematics at all,' recalled Minowsky, one of Einstein's teachers at the Zurich Technical Hochschule. Unless we take into account the profound alignment of Einstein's chart to the grand opposition taking place in the sky, we will be hard put to see why he in particular created the new synthesis. Ever since 1887, when experiments by Michaelson and Morley supposedly failed to find any 'ether-drift' (i.e. they found no difference in the speed of light in different directions in relation to the Earth's movement through space), all the necessary ingredients for the new synthesis had been available.

The orientation of Albert Einstein was to the outermost spheres, beyond the merely personal. 'A mind profoundly self-enclosed and remote' as Lord Keynes said of Newton, applies equally well to him. Einstein was asked to compose an autobiographical statement.[14] This stated: 'Even when I was a fairly precocious young man, the nothingness of the hopes and strivings which chase most men restlessly through life came to my consciousness with considerable vitality ...' He described his endeavour to free himself from the 'merely personal', by contemplation of a non-human external world. Half of this autobiography is devoted to equations.

The chart of Albert Einstein (figure 11.3) shows the 'quintiles of creativity' as symmetric about a Jupiter-Uranus axis. The chart depicts only quintiles and biquintiles aspects, with orbs set rather wide to show this symmetry, plus one opposition. It may help, while mulling over the exquisite symmetry of this chart, to have a quote about Einstein's view of the world:

> *'To know that what is impenetrable to us really exists, manifesting itself as the highest wisdom and the most radiant beauty, which all our dull faculties can comprehend only in their most primitive form, — this knowledge, this feeling, is at the centre of true religiousness.'*
>
> Einstein, *The World as I see It*

The lunar nodes crossed over the axis of Einstein's chart during the 1905 period of inspiration, as too did Saturn. Uranus and Neptune formed a trine roughly symmetric about the ascendant. The theory dawned upon him in 1905 as Neptune had moved round a sextile to meet his ascendant, while Uranus had moved round a trine to meet his descendant (one moves at twice the speed of the other).

A diagram of the Einstein transits through the formative year 1905 (figure 11.4) shows the triple meeting of his Uranus-trine return, shadowed by the slower Neptune-sextile return. The figure shows these transits out to only one degree of orb, whereas a somewhat wider scope, say three degrees, is usual. It depicts how the Uranus trine became exact thrice that year, as is normal for outer-planet transits, and how his Neptune-sextile return was shadowing it. This helps us to appreciate how the Uranus-Neptune opposition then occurring was linked to his own chart. An opposition of Saturn to his natal Uranus is also shown.

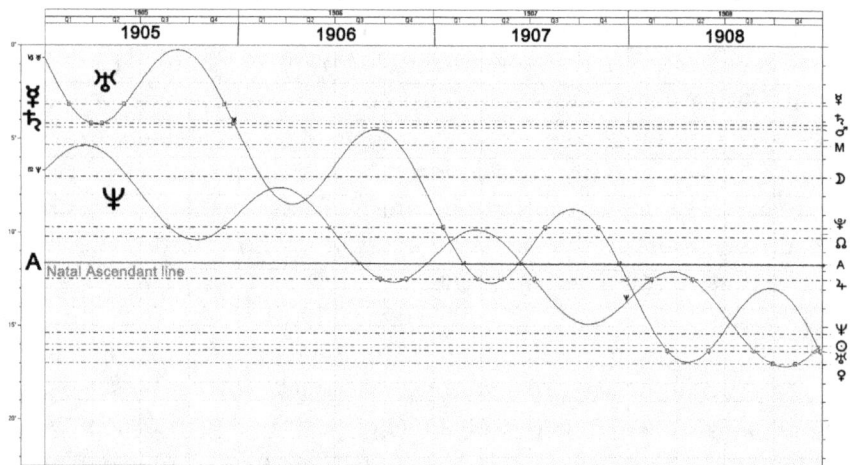

Figure 11.4: The Neptune-Uranus opposition of 1905 shown in '16th Harmonic' mode i.e. 22.5°.

The opposition between Uranus and Neptune kept meeting and re-meeting for almost a dozen times over five years or so, owing to their retrogradations. On their first encounter around March 17th, 1905, they approached within a degree of each other. This was the date on which Einstein completed his paper about what he called the 'photon' which was the quantum of light, and for which he was later

Chapter 11. Destiny and the Uranus Cycle

awarded the Nobel Prize. The great opposition in the heavens was in square to Saturn and Mercury in Einstein's chart, as were closely conjunct. Over the next year or two, 1906–7, the opposition kept criss-crossing the Einstein ascendant.

In April of that year, he finished his PhD and submitted it, and sent off his paper introducing the 'photon' of light. Then some weeks later, his relativity paper was finished at the second of the Uranus-trine returns. The third Uranus-trine return chimed as he was formulating his historic equation, $E = mc^2$.** The paper containing it was received by the *Annalen der Physik* on the 27th of September, 1905.

The equation was composed a month or so prior to that date, so all we can say is that the figure indicates some major transits leading up to it. Saturn first met his Jupiter in March, then moving retrograde again met it in September, presumably as the historic equation was being composed.

The slow meeting of the two outer spheres is diagrammed by figure 11.4. The horizontal lines show certain features of the Einstein chart: his ascendant, plus his Saturn conjunct Mercury at right angles to it. This is a fourth-harmonic chart in that it extends over 90° and shows conjunctions, oppositions and squares all as conjunctions. (A 16th harmonic here used may show more clearly the planetary motions.) In the spring of 1905, as the opposition came within a degree of exactitude, being then in square to his Saturn and Mercury, Neptune was spot-on Einstein's natal Uranus-Neptune midpoint, at four and a half degrees of Cancer. Later in the year it started crossing over his ascendant.

The Saturn conjunct Mercury in his chart points to his slow but thorough mental activity, plus 'a propensity to solitary habits':[15]

> When I asked myself,' he confided to a friend, 'how it happened that I in particular discovered the Relativity Theory, it seemed to lie in the following circumstance. The normal adult never bothers his head about space-time problems. Everything there is to be thought about,

---

**One expert, H. J. Haubold, advised the writer that the idea of $E = mc^2$ was formulated 'somewhere in between' the 30th June paper and its presentation in the 27th September paper, of 1905. There is no more definite information available.

in his opinion, has already been done in early childhood. I, on the contrary, developed so slowly that I only began to wonder about space and time when I was already grown up. In consequence I probed deeper into the problem than an ordinary child would have done. *Koestler, AOC, p.175.*

Historians have not had a lot more than this to go on, as regards the springs of Einstein's creative power — and that isn't much. Earlier, he had been 'a backward child, a slow developer, a drop-out from school'.[16]

Saturn was crossing his chart's major axis, having just formed an opposition to his natal Uranus. That Saturn was in trine and sextile to the ongoing Uranus-Neptune opposition, and conjunct the lunar-node axis, as both were passing over his chart's Uranus. The paper he produced contained no references nor did it acknowledge anyone, except his friend Besso, also working in the Patent Office: it was a supreme example of Pure Thought.

On June 30th, this paper on Special Relativity was received by the *Annalen der Physik*, as a result of that insight somewhere around May 20th, when the Einstein transits were:

Neptune	sextile	Neptune	1°
Uranus	trine	Uranus	1°
Jupiter	square	Jupiter	1°
Moon	opposition	Moon	2°

In addition, he had the North Node conjoining his Uranus to half a degree, Saturn (conjunct the South Node) opposing that Uranus within 1°, and Uranus squaring his natal Saturn to 2°. He was then 26 years old, when Neptune had gone round almost one-sixth of its huge orbit (its sextile return), and Uranus one-third of its 84-year cycle (its trine return). These transits are impressive, and appropriate for the event.

The nodes have always been regarded as powerful parts of the ecliptic, as comprising the axis along which the lunar orbit intersects the plane of the ecliptic, and are linked to the notion of fate and destiny. They are the points on the zodiac where eclipses happen. To have the nodes exactly aligned with such a potent axis of symmetry in one's chart, on the day one's theory is completed, is destiny in no

Chapter 11. Destiny and the Uranus Cycle

uncertain terms. Six weeks earlier at his E-moment it was getting close, being four degrees away.

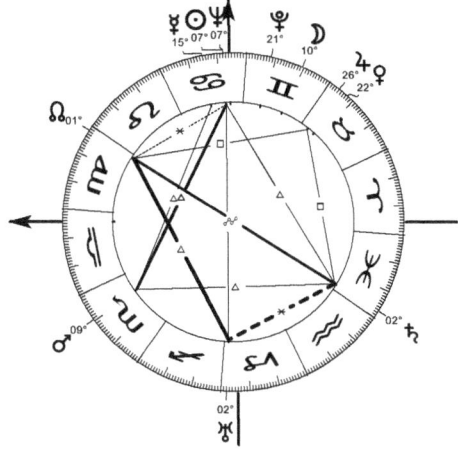

Figure 11.5: Event Chart for Relativity Theory received, showing 'Mystic Rectangle', 30 Jun 1905 12:00hrs midday (−01:00) CET, Berlin. (7° orbs and aspects to node used.)

The chart of June 30th (figure 11.5) shows a 'mystic rectangle,' as could symbolise the four dimensions of space-time in the new theory, aligned both to Einstein's ascendant and to the 'main axis' Jupiter-Uranus of his chart. With rather wide orbs, it also shows a grand trine aligned to his chart. Einstein had Pluto square to his Jupiter, and on June 30th, Jupiter and Venus were (within 4°) conjunct over that Pluto. Also, the Sun (and Mercury) were chiming in conjunction with Neptune, and together they transited the Einstein ascendant (3°). In other words, they were together occupying the same part of the zodiac, as had been rising when Einstein had been born. To have the Sun and Neptune jointly cross one's ascendant is an illuminating and somewhat transcendental period.

Those are some of the more remarkable points in a compelling alignment between a natal chart and a gestation-process for a theory. It may here not matter that we lack the exact day for the realisation, as the planets that really matter here are the outer slow-moving ones. The transits are more exact and notable at the time when the theory was presented to the world, than at the more subtle and private moment of realisation.

At the mysterious heart of the scientific creative process are the prime-number aspects five and seven, the quintiles and septiles, clashing with the traditional symmetries of the zodiac, its twelvefold structure. Uranus and Neptune have quintile and septile aspects in Einstein's chart: Neptune with two quintiles, as mentioned; and Uranus, a quintile and a septile. The eureka-date of May 20th (or thereabouts) was strong in both of these aspects.

In 1907, Einstein 'made the first important strides towards the general theory of relativity',[17] that linked gravity to his speed-of-light equations. This year was at the centre of the great opposition, while it was criss-crossing his own ascendant. Synchronously, the modern art movement of Cubism began in this year, though apparently not consciously influenced by the new abstractionism in space and time that physicists were coming up with.[18]

Fame

A test of the Theory of Relativity was made at a solar eclipse in May 1919. This revealed that light rays bent slightly as they went past the Sun. This conclusion was announced at a joint Royal Society-Royal Astronomical Society meeting on November 6th, 'the day on which Einstein was canonised'[19] in the words of his biographer. Uranus went stationary (i.e. ceased its apparent motion in the sky) on the 5th–6th. Einstein's natal Jupiter was at 27° 29' of Aquarius, while Uranus was stationary at 27° 46' of Aquarius: a conjunction of scientific precision. The nodes at 25° Scorpio were then conjunct Einstein's Pluto, in square to his Jupiter.

A square pattern built up in the heavens in relation to the natal chart, on the day when Einstein's theory became firmly established.

REVOLUTION IN SCIENCE
NEW THEORY OF THE UNIVERSE
NEWTONIAN IDEAS OVERTHROWN

trumpeted The Times of November 7th, 1919.

Space 'warped',

noted a laconic subheading. There was a lunar eclipse that day. Newspapers had to grapple with the fourth dimension, as Uranus reached the axis of Einstein's chart, opposite its natal position. Jupiter is here appropriate in terms of fame and recognition arriving.

'Press announcements such as these mark the beginning of the perception by the general public of Einstein as a world figure' to quote Einstein's biographer. For 'the suddenly famous Mr Einstein', his Uranus-opposition marked his emergence into the glare of publicity, the zenith of his career — a powerful vindication of the Tarnas thesis.[20] His confidence is well shown by an anecdote from this period: asked what would happen should the experiment fail to confirm his theory, Einstein replied: 'Then I should feel sorry for the dear Lord. The theory is correct.' A mass meeting was arranged against general relativity theory in Berlin in August 1921, which Einstein attended (Uranus opposition Uranus $2\frac{1}{2}°$). He sat in a box and is said to have enjoyed the occasion. At the beginning of this, we described inspiration dawning upon Einstein at his trine position, and that reached fulfilment at the opposition: a 60° angle between them. Thereafter, despite every encouragement, Einstein accomplished little by way of scientific work.

Einstein wrote a letter to President Roosevelt, urging that the atom bomb be constructed. This letter was delivered to Roosevelt on 11th September 1939, and its effect was decisive[21] — to Einstein's undying regret in later years. Mars was then at 26° 14′ of Capricorn, within arc-minutes of his natal Mars.

Urania

Let us recapitulate on these themes. When Newton died, Uranus had completed one orbit, and Neptune just a half, so that his transits were:

| Neptune | opposition | Neptune | 3° 05′ |
| Uranus | conjunct | Uranus | 1° 21′ |

Similar comments apply to the great astronomer Sir William Herschel who discovered Uranus. His death-transits in 1822 were:

Neptune	opposition	Neptune	0° 15′
Uranus	conjunct	Uranus	0° 59′
Node	opposition	Node	0° 18′

Those are close, very close. Are they trying to tell us something? The nodal axis was lined up with its position at his birth, having made four and a half revolutions (each of 18.61 years) in his life of 84 years. These periods display harmonic ratios between the orbit periods in our solar system.[22] They are inter-related via that of Uranus. Sir William Herschel was a musician and a conductor, interested in the theory of music as well as in deeper matters of Pythagorean philosophy, so he might have appreciated the following: twice the Uranus orbital period (83.75 years) is close to the Neptune orbit of 164.1 years, thrice to the Pluto period of 247.7 years, one-third to the Saturn period of 29.46 years, and one-seventh to the Jupiter period of 11.9 years.[††] Its zodiacal revolution of 84.01 years is rather precisely the product of 12 and 7, numbers which we are hardly at liberty to regard as unimportant.

Of those ratios, the 3:1 ratio between Saturn and Uranus is not exact, so that one cannot experience the full Uranus return with a Saturn return. They do however interact in the late twenties, with the Uranus trine at around 28 years and the Saturn return at $29\frac{1}{2}$ years. Astrologers refer to the three outer planets as 'transpersonal', yet they measure out vital life-stages. These orbit periods are interrelated precisely enough to coincide in the transits measuring out human life-cycles.

For astronomers, it is a mere anomaly that Neptune turned up where it was not expected and thereby dissolved the credibility of Titius-Bode's Law. This was an empirical law defining relative distances between the planets, and was working quite well even for the position of the asteroid belt, until it predicted that the next planet beyond Uranus should be positioned where Pluto was eventually found. 'Very Neptunian', astrologers may mutter. From our point of view, it is no mere anomaly but a vital property of the *Harmonices Mundi* that Kepler attempted to formulate, that Neptune should have double the orbit period of Uranus.

Neptune's long cycle can have a personal significance for an inventor. Einstein invented his Theory of Relativity as his Neptune-sextile and Uranus-trine returns chimed together, within a degree. We saw

---

[††] One-ninth is close to the time that the nodal axis takes to return to its original position, i.e. half its rotation period of 18.61 years, as shown in the transits at Herschel's death.

Chapter 11. Destiny and the Uranus Cycle

how anaesthesia was discovered in the weeks following Neptune's discovery: William Morton, the Boston dentist, sent himself into unconsciousness with ether on September 30th, 1846. At twenty-seven years of age he had arrived at his Neptune sextile, within a mere 16′. Saturn and Neptune were conjunct, remaining so on October 16th, when Morton initiated the practice of surgery under anaesthesia (described in chapter 9: When Inventions Worked).

Sigmund Freud died at his Uranus-return, and Tarnas obtained his theory about Prometheus/Uranus through studying the life of Freud. His death-transits were:

Jupiter	conjunct	Jupiter	4° 44′
Uranus	conjunct	Uranus	1° 06′
Neptune	opposition	Neptune	3° 28′
Pluto	square	Pluto	1° 53′

Freud's life measured out seven Jupiter cycles, one Uranus cycle and half of a Neptune cycle.‡‡ These do seem appropriate for such a life. Tarnas concluded that Freud's most creative period came in the latter half of 1895, over his Uranus-opposition (and Neptune-square) return, when his theory of dreams came to him. It is evident that Tarnas' theory has far-reaching repercussions for scientific biography.

We conclude with three further instances:

- **Ellipses dawn on Kepler:** The most important discovery that Kepler ever made dawned on his Uranus-trine return. In 'early April' of 1602[23], he finally came to appreciate that the orbit of Mars could not be circular, as had been assumed for two thousand years, but was elliptical. This was at Uranus trine Uranus (orb of 0° 50′). Thereby, he was able to formulate the first of his three immortal laws. Curiously enough, this was *after* his Saturn return, which arrived at the end of 1601. Kepler had Sun conjunct Uranus in his natal chart.[24]

- **Leibniz envisions Calculus:** In the archives of Hannover are the precious manuscripts on which Baron Gottfried von Leibniz' invention of the calculus is recorded, dated 25th–29th

---

‡‡As additional chords of the final note, Freud also managed to die on a Saturn sextile Saturn (2° 21′), Mars trine Mars (3° 33′) and Mercury trine Mercury (2° 33′) returns.

October 1675. He, there, first used the ∫ and dy/dx symbols for integration and differentiation, which now, three centuries later, are in use worldwide. That genesis-moment, the grand insight of his life, fell within 0° 13′ of his Uranus-trine return.

- **Darwin and Natural Selection:** Uranus and Neptune cycles interlinked at Darwin's trine-return, (figure 11.6) as the theory of evolution dawned upon him. Uranus transiting his natal Mercury is also shown. It is thereby evident that Darwin had a Uranus trine Mercury in his natal chart. Dr Tarnas classifies Darwin as a Mercury-Uranus type. Noting that he was born on the same day as Abraham Lincoln (12th February, 1809), Tarnas observed that both were 'distinguished for their intellectual brilliance and independence as well as their powers of communication in the service of major revolutions'.[25]

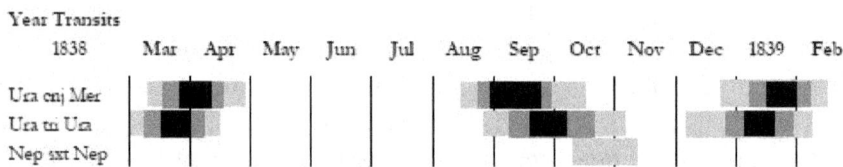

Figure 11.6: Darwin's Uranus-trine Return — 1838 Year transits.

- **The Rutherford atom:** On May 11th, 1911, while teaching at Manchester University, Ernest Rutherford's paper, 'Theory of the Structure of the Atom' was published in the *Philosophical Magazine*. He had envisaged the modern picture of the atom, as having a small, dense nucleus at its centre. His Uranus-opposition return then stood at 0° 4′. The idea came to him at an unspecified date in December of 1910, as his Uranus return was coming into focus. Rutherford later matured into the 'grand old man' of British science and President of the Royal Society.

In the lives of scientists, two events often intertwine around a highly creative period in life: Saturn more for establishing one's life-direction while the Uranus trine represents more the creative insight. The latter aspect always occurs as a threefold process, for astronomical reasons, and it was the last of these that chimed when Kepler

Chapter 11. Destiny and the Uranus Cycle

discovered elliptical motion. The theory advanced by US philosopher Dr Richard Tarnas has here been viewed as a key to scientific biography, as concerning what one might call the geometry of self-realisation.

****

Some readers will, one hopes, be uneasy at the ascribing of significance to transit angles: say, a 90° angle between Jupiter at Einstein's birth, and its position in the sky at the time of completing his theory. How can an angle be measured from a position which no longer exists? Where is it remembered? On what ethereal tablets is it inscribed? For an answer we may go back some centuries earlier to what Kepler wrote.[26] As the following texts are hardly available in English, we may quote them at some length. It was his view that:

> When a human being's life is first ignited, when he now has his own life, and can no longer remain in the womb — then he receives a character and an imprint of all the celestial configurations (or the images of the rays intersecting on Earth), and retains them unto his grave ... This character is not received in the body, which is much too ungainly for that, but rather in the nature of the soul itself, which is like a point. The soul can therefore transform herself in that point at which the rays converge, and can not only partake of that faculty of reason for which we alone among living creatures are called reasonable; she is also implanted with another kind of reason — geometry, which can be perceived instantaneously in the rays as well as in the sound of music without a lengthy learning process.
>
> *Tertium Interveniens*, 1610.

and that:

> The vital power that is ignited in the heart and burns there as long as life lasts is, in a certain sense, a zodiac. It consists of energy and of a fiery outpouring. Thus the whole sensually perceptible figure of the zodiac flows

> into the vital functions as soon as they are ignited at birth, and grows into them completely ... At that point at which it [the vital power] begins to be what it was then [at birth], when it incorporated the harmonies into itself, then the sensually perceptible harmony of the planets' rays flows into it most strongly.
>
> Harmonices Mundi, 1619.

From which it followed:

> In the vital power of the human being that is ignited at birth there glows that remembered image to which I referred also in connection with the Earth-Soul. So great is the persistence of the celestial character-image and of all the details of the natal theme, and so durable is the natal image of the soul, that it is not forgotten until life's end. With all planetary transits over the more significant positions of the horoscope, the above-mentioned vital power is aroused, just as if those positions were not mere images of long-past things remaining in the soul, but rather real celestial bodies. Thus, for example, it is as if there were not one, but two suns in the sky which are unified into one, and which set into action, by their union, the nature of the vital power.
>
> Ibid, as translated by Negus, 1987.

These things were not to be taken in a fatalistic sense, Kepler advised, explaining that:

> The birth constellation is merely an empty form. When the soul is poured into it, the human being does not take on this form immediately. He is unwieldy, maintains much from the thinking and feelings of the mother and mixes that with the nativity. Thus the nativity forms and shapes the soul, but it does not bring forth a new one, nor does it change the soul completely.
>
> letter to Fabricius.

Chapter 11. Destiny and the Uranus Cycle

Relevant to invention-moment charts is his comment (written during his Saturn return) that:

> *that power that makes the aspects effective must be inherent to all sublunar bodies, indeed the whole earth. The entire vital power is, you see, a reflection of God, who creates according to geometric principles, and is activated by this very geometry or harmony of the celestial aspects.*
>
> De Fundamentis, 1602.

The British took Kepler as an astronomer, and viewed what they saw as his other side as 'mystical'. There was, however, nothing mystical about Kepler. These quotes show the practical and realistic view he took towards these difficult matters. These comments upon the how of celestial influence have scarcely been translated at all into English over the centuries. Only some partial versions of Kepler's astronomical texts have been so translated. These texts can only be quoted at all because Professor Ken Negus has translated and published some relevant passages.[27]

References

1. Zadkiel, *The Grammar of Astrology*, 1849.

2. W. Simmonites, *Complete Arcana of Astral Philosophy* 1890; L. Broughton, *The Elements of Astrology*, 1898, New York. These sources were kindly found by Clive Kavan from his library.

3. R. Tarnas, 'Uranus and Promethius' June 1990, *The Astrological Journal* p.150. (Reprinted in his *Prometheus, the Awakener*, 1995.)

4. S. Freud, preface to 'The Interpretation of Dreams' 3rd Edn., quoted by Tarnas in *Prometheus the Awakener, An Essay on the Archetypal Meaning of the Planet Uranus* 1995, Spring Publications CT, p.33. This is my favourite astrological text! Dr Tarnas magnum opus is *The Passion of the Western Mind, Understanding the Ideas that Have Shaped our World View*, (Yale U.P. 1992).

5. Tarnas, op. cit. (3), p.34.

6. N.K., 'The Uranus Cycle in Isaac Newton's Life', *The Mountain Astrologer* April 2000, pp.40–45; N.K., 'Newton, Halley and Uranus,' *The Astrological Journal*, July 2002, pp.23–28.

7. R. Westfall, *Never at Rest*, a biography of Isaac Newton, 1980, pp.134,140.

8. Ibid, p.606.

9. N.K., 'Charts that Changed the World' *The Mountain Astrologer* Oct 1993; also, N.K., 'Revolutionary Innovators and the Uranus Cycle' *Geocosmic Magasine* ISAR US Winter 1995/6.

10. S. Chandrasekhar, *Truth and Beauty, Aesthetics and Motivations in Science* 1987, p.60.

11. A. Pais, *'Subtle is the Lord'*, *the science and life of Albert Einstein*, Oxford 1982, p.139.

12. Letter from H. J. Haubold, Akademie der Wissenschaften der DDR.

13. Ibid.

14. Albert Einstein, *Autobiographical Notes, A Centennial Edition*, 1979 Trans. & Edited by P. A. Schilpp.

15. Charles Carter, *The Astrological Aspects*, 1930 p.105.

16. Brian Inglis, *The Unknown Guest, The Mystery of Intuition*, 1987 p.88.

17. Pais, ref. (11), p.48.

18. Leonard Shlain, *Art and Physics, Parallel Visions in Space, Time and Light*, US 1991, p.203; also Arthur Miller, *Insights of Genius, Imagery and Creativity in Science & Art*, 1996.

References

19. Pais, ref. (11), p.395.

20. Tarnas, op. cit. (3) p.99.

21. Richard Rhodes, *The Making of the Atomic Bomb*, 1986, p.313.

22. John Martineau, *A Book of Coincidence* 1996; *A Little Book of Coincidence in the Solar System*, 2006.

23. W. Donahue, 'Kepler's First Thoughts on Oval Orbits' *Journal for the History of Astronomy*, 1993 xxiv pp.71-10,73.

24. N.K., 'Kepler's Chart', *The Astrological Journal*, Nov. 1996, pp.371-7.

25. Tarnas op. cit. (3) p.58.

26. N.K, 'Kepler's Belief in Astrology' in *History and Astrology, Clio and Urania Confer* Ed. Kitson, 1988.

27. Ken Negus, *Kepler's Astrology, Excerpts* Princeton, 1987. Prof. Negus, a German scholar at Princeton University, founded the Princeton Astrological Association. N.K. and Nick Campion, 'Kepler's Astrology', emphCosmos & Culture, Autumn 2011 special issue.

CHAPTER 12

DNA and the Decile

> '...harmonic patterns (3 or more planets in a harmonic) are the single most important astrological influence in the chart.'
>
> David Cochrane.[1]

The quality of a moment of time is related to the 'harmonic then dominant'. We saw how this term pertains to celestial aspects, and that e.g., the fifth harmonic embraced both quintile and biquintile aspects. But, what would be the point of using e.g., a 'tenth harmonic', as would mean slicing up the circle into ten pizza sectors? Wouldn't it be easier just to stick to deciles, of thirty-six degrees? An answer will here be given, with the aid of a computer graphic technique, as depicts the intensity of a given harmonic flowing through time.

The harmogram concept has here been introduced, to show how a given harmonic varies over some time-interval. Whereas astrology has always focused on the single moment of time, when the chart was cast, we can now at last behold the strength of an aspect-pattern as a *flowing quantity* over any given period. As calculus liberated mathematics from fixed geometric diagrams, by introducing time-varying functions, so likewise the harmogram liberates astrologers from the limitation of a fixed moment of time.

Let us look, as astrologers do, at the moments of birth — of two astronomers, of a mathematical theory, a new element, and a chemical model. Two of the events have already been discussed. At these formative moments, we examine the strength of the celestial aspects then present.

Tycho Brahe and Kepler: 5 and 7

In Benatek castle outside Prague, Tycho Brahe and Johannes Kepler spent twenty tempestuous months together, whereby the foundations of modern science were laid. It is hard to imagine two people sitting at a table with greater disagreements: for the one, the earth was immovable and at the centre of the universe, while for the other it had a triple motion and it was the Sun that stood still. For one, all of the stars in heaven whirled around the *axis mundi* every day, while for the other they stood still. Kepler had not yet reached his theory of elliptical motion during this collaboration.

Kepler described his noble colleague as one whom it was impossible to approach without exposing oneself to the gravest of insults. Had Brahe not died just when he did, Kepler would probably have ridden off after one more row and never got his hands on the priceless data. It was understandable that Brahe should be in a bad mood: his beloved Uraniborg on the island of Hveen off Denmark, where he had initiated modern astronomy by taking the first systematic observations of the heavens, was falling to mere rack and ruin owing to a change in the Danish monarch (Something rotten in the state of Denmark, perhaps?). Brahe now held the post of Imperial Mathematician of Prague, doubtless a fine post, but the person destined to be his successor and print all his data remained radically sceptical about Brahe's scheme of things.

Brahe had seven septiles in his chart (using the 'Addey orb'), which perhaps assisted him in a new scheme of the universe which he designed. He readjusted the traditional seven spheres of the heavens. It is easy to forget the major role which Tycho Brahe's view played, and see the transition as merely from Ptolemy's geocentric scheme to the Copernican. Brahe's view appeared as a sensible compromise.

Let us hope that Kepler refrained from making sarcastic remarks about how the orbits of Mars and the Sun intersected in Tycho's scheme of things. Brahe conferred upon all the planets except Earth an orbit around the Sun, and then made the Sun revolve around Earth.

Chapter 12. DNA and the Decile

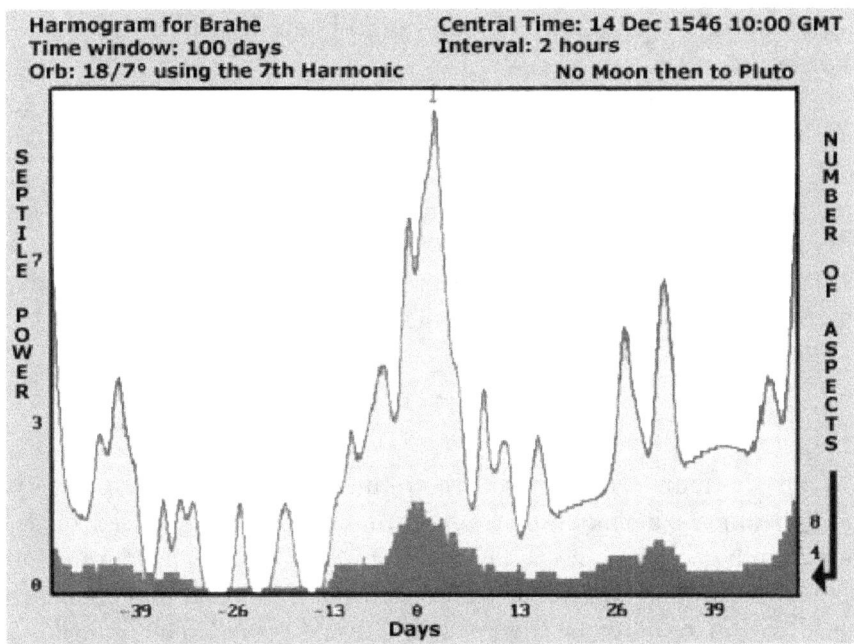

Figure 12.1: 7th harmonic harmogram For Tycho Brahe's Birth ±50 days.

The two astronomers shared the conviction that the planets influenced life on Earth, or that as Kepler put it, there is nothing that happens in the heavens that is not sensed in some manner by the Earth below. They provided no clearer evidence for that affirmation, than the timing of their own births.

Taking a hundred-day period, centred on Brahe's birth, the computer scores the septile aspects, as a continuous time-function. The harmogram (figure 12.1) shows a huge peak just before Brahe's birth (it peaked about twelve hours before). So what? Perhaps in this case, the seven septiles helped him to re-envisage the total scheme of things. The graph plots 'septile power', based on a concept which Tycho Brahe would have understood: that an aspect is stronger when nearer to exactitude. The computer scores each aspect in proportion to its closeness to exactitude and then sums them. So, if a number of septiles appear which are all rather weak, i.e. of wide orbs, then they will score low on the 'septile power' graph.* At the base of the

---

*For the 7th harmonic harmogram centred on Brahe's birth, the program was set to discount lunar septiles. These come and go more quickly than the planetary

263

harmogram is a bar-chart of the score of septile aspects present in the sky.

Accounts of the Danish noble Tycho Brahe tend to focus on his quirky behaviour: his brass nose, his pet elk and his psychic dwarf, and how he ruled over the isle of Hveen with the aid of a prison he put people in. Rather, we should respect those who were the last of a tradition whereby the heavens were to be interpreted as well as observed. For, after Brahe and Kepler, the abyss opened up, so that instead of meaning, astronomers merely purveyed information.

****

For the man who invented the quintile, it is appropriate to display a quintile harmogram (see figure 12.3 overleaf), based on the pentagram geometry of the number five. A close-up harmogram (not shown) of the day of Kepler's birth indicates how his birth synchronised within *minutes* to the peak in quintile power! The quintiles in his chart formed a triangle (figure 12.2), as seems quite appropriate for a *mathematicus*.

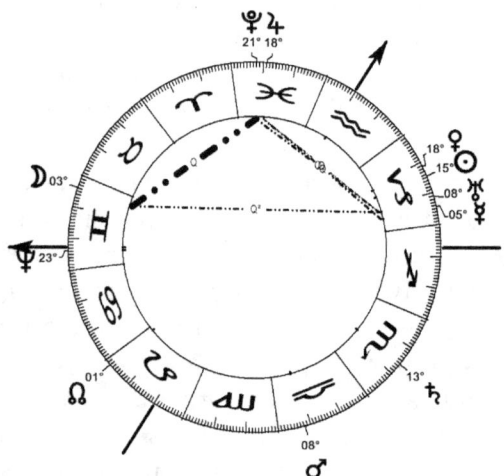

Figure 12.2: Natal Chart for Kepler showing quintiles only, 27 Dec 1571 OS 14:30hrs LMT, Weil der Stadt, Germany. (Addey orbs used.)

---

aspects and would make such a long-period harmogram rather blurred.

Chapter 12. DNA and the Decile

Thus, Time has inscribed indelibly the numbers Five and Seven upon the fates of these two, a couple without whom the scientific enterprise would have been unthinkably different. Without Brahe's systematic observations there would have been no Kepler's laws, no physical image of how the solar system worked for the first time, and without that there would have been no Newton's *Principia*. It is worth pondering the significance of these mould-breaking prime numbers in that historic context. Let's here cite a very early (perhaps the first) account of what a quintile did, as involved the weather.

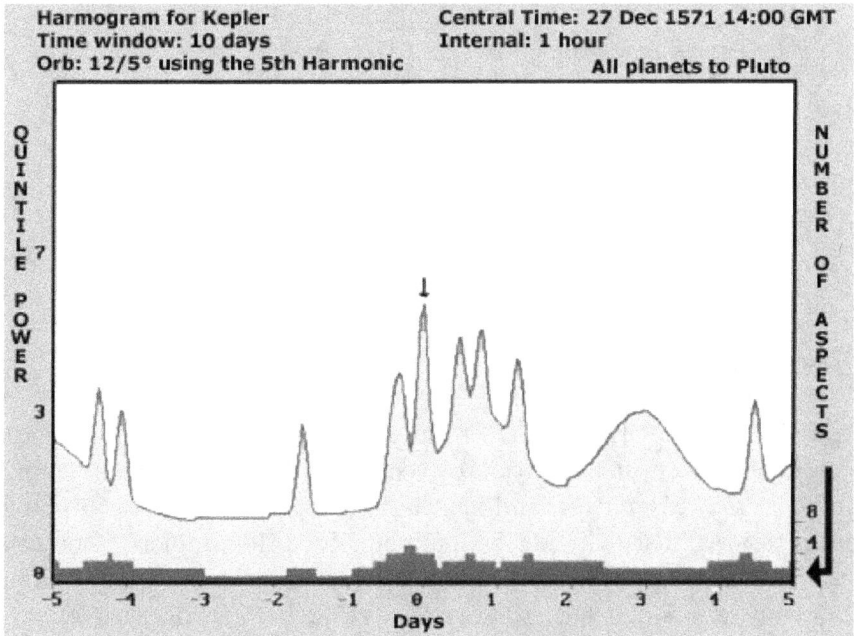

Figure 12.3: 5th harmonic harmogram For Kepler's Birth.

One of Kepler's correspondents, the physician Georg Brengger, had asked him how anyone could make a weather prediction based upon planetary aspects. Were there not too many of them to be able to discern what was what? Kepler published a yearly almanac, printing a modest 500 or so, giving prognostications for the coming year on matters of general concern — abundance or famine from the harvest, political turmoil, weather, etc. — and was concerned to explain equally the rational basis thereof and the limits of what could be foreseen. 'You say you have confirmed by meteorological experience that there exist the additional aspects quintile, biquintile and sesquiquadrate. I myself should like to see an example of this observational material ...' pertinently queried Brengger. Kepler's reply cited an example as requested:

> In 1600, when from 23 April until 2 May, New Style, there were no primary aspects, and Magini's tables showed Saturn and Jupiter to be at quintile, on 1 May there was a very heavy fall of snow both in Prague and in Styria for Ferdinand's wedding, and the jousting had to be cancelled. From observing the heavens, it was found that during these same days Saturn and Jupiter were 72° apart. Tycho's students made the check on my behalf with Tycho's quadrant.[2]

What may be the first evidence for the effect of a quintile involved the cancelling of a jousting tournament!

The Quaternions of William Hamilton

In 1843, William Hamilton, President of the Irish Royal Academy, was trying to effect a rotation of vectors in three-dimensions. Long he pondered this matter, and his children used to ask him as he came down to breakfast, whether he had yet solved the problem. One day, as Hamilton was walking to a Council meeting of the Royal Irish Academy, and as he was crossing over Brougham Bridge, the answer came to him.

Suddenly, he realised that *four* separate components were required, not three as he had all along supposed. The operators, which he was to call 'quaternions', had a fourfold structure to them.

To quote his biographer:

> This instance of scientific creativity, following on long digestion of a problem, is certainly one of the best documented in all science history. Adequate information exists about the general background as well as the precise manner in which sudden inspiration came. As such, it

Chapter 12. DNA and the Decile

remains an episode of outstanding interest to students of creativity, as well as to scientists generally.

For this, we can indeed be grateful. However, students of creativity will have to look in an unaccustomed direction if they wish to comprehend the reason for this flash of inspiration, namely the heavens above.

Pausing awhile, the President pulled out his pocket knife and engraved upon Brougham Bridge the formula which had then struck him. No trace remains of the engraving, but who would want to doubt such a story? Perhaps his wife Helen who was accompanying him found this rather peculiar. Today a plaque on the bridge marks the spot, which reads:

> Here as he walked by
> on the 16th of October 1843
> Sir William Rowan Hamilton
> in a flash of genius discovered
> the fundamental formula for
> quaternion multiplication
> $$i^2 = j^2 = k^2 = ijk = -1$$
> & cut it on a stone of this bridge.

One hopes the reader will not expect elucidation of this deep insight from the present writer. It is said to express the theoretical principle of the Rubik cube.

The next day, Hamilton wrote to a colleague describing his new insight. Regrettably, fifteen years were to go by before an account of this eureka moment was vouchsafed to anyone. Hamilton's disciple, P. G. Tait, was told the following:

> *October 15th, 1858*: (Quaternions) started into life, or light, full grown, on the 16th of October, 1843, as I was walking with Lady Hamilton to Dublin, and came up to Brougham Bridge ... That is to say, I then and there felt the galvanic circuit of thought *close*; and the sparks which fell from it were the *fundamental equations between i, j, k, exactly such as I have used them ever since.*
>
> I pulled out, on the spot, a pocket-book which still exists and made an entry on which, *at that very moment*, I felt

that it might be worth my while to expend the labour of at least ten (and it might be fifteen) years to come.

One month before his death, Hamilton again recorded the same event for the benefit of his son Archibald and likely future psychologists. Though the words now are different, the meaning is similar:

> Although (your mother) talked with me now and then, yet an *undercurrent* of thought was going on in my mind, which gave at last a result, whereof it is not too much to say that I felt *at once* an importance.

> An *electric* circuit seemed to close; and a spark flashed forth, the herald (as I *foresaw, immediately*) of many long years to come of definitely directed thought and work ... Nor could I resist the impulse — unphilosophical as it may have been — to cut with a knife on a stone of Brougham Bridge, as we passed it, the fundamental formula ...[3]

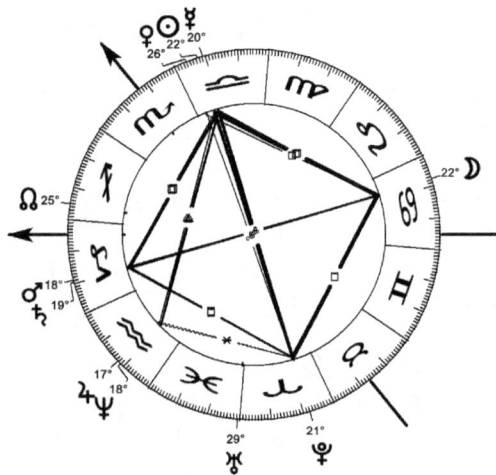

Figure 12.4: Event Chart for Quaternions, 16 Oct 1843 NS 13:05hrs LMT, Dublin, Ireland. (5° orbs used.)

It seems that Sir William Hamilton did indeed spend the next twenty years of his life elucidating the meaning of that insight which came to him on the bridge, on Monday, 16th October 1843. 'Quaternions were the gateway to modern algebra' his biographer explained, and led to a flurry of activity amongst British mathematicians. To this day, Hamilton's name remains the most celebrated amongst Irish scientists and mathematicians.

At that moment in time, the chart of the event (figure 12.4) shows most of the solar system configured into a square (a 'grand cross'). If anyone wants concrete proof for astrology, here it is. The harmogram of this moment (figure 12.5) depicts the 'fourth harmonic': that is to say, it measures the strength of square-plus-opposition aspects forming on the ecliptic at that historic noon of October 16, 1843. The odds against such a combination turning up by chance are astronomical. Only three months before this moment, was there a bigger peak, in this harmonic. It is clear that the heavens predisposed the mind of Ireland's greatest mathematician to receive the answer he was looking for, at a particular moment in time.

The peak of the harmogram occurred at noon of that day. Can we estimate when Sir William Hamilton would have been crossing the bridge? There was a meeting of the Royal Irish Academy at 3.30 p.m. that day, and a Committee of Council meeting before it, at 3.00 p.m. Presumably as President he would have been expected at the earlier meeting. The mathematics professor A. G. O'Farrell recently re-did the historic walk on the century and a half anniversary of the event. He found that from the bridge it took about 25 minutes to a nearby village to get a taxi, which would take another 15–20 minutes driving to Dublin. Yet Hamilton's account said he was walking to Dublin, which would have taken say an hour and a half. This gives us a conjectural time of around 1 p.m. when he was on the bridge local time, or 13:30hrs GMT. At the meeting that afternoon, Hamilton announced that he had reached an important new insight, and requested that he be allowed to present it the next day.

In a letter to the writer, Professor O'Farrell added cryptically, 'Incidentally, I have always assumed that he walked under Brougham Bridge'. The picture of this historic bridge given in Hamilton's biography has water flowing under it. Clearly, eureka researchers will wish to resolve this matter.

The investigation of this mathematical E-moment was suggested by science historian, B.C. The policy of avoiding mathematical E-

EUREKA

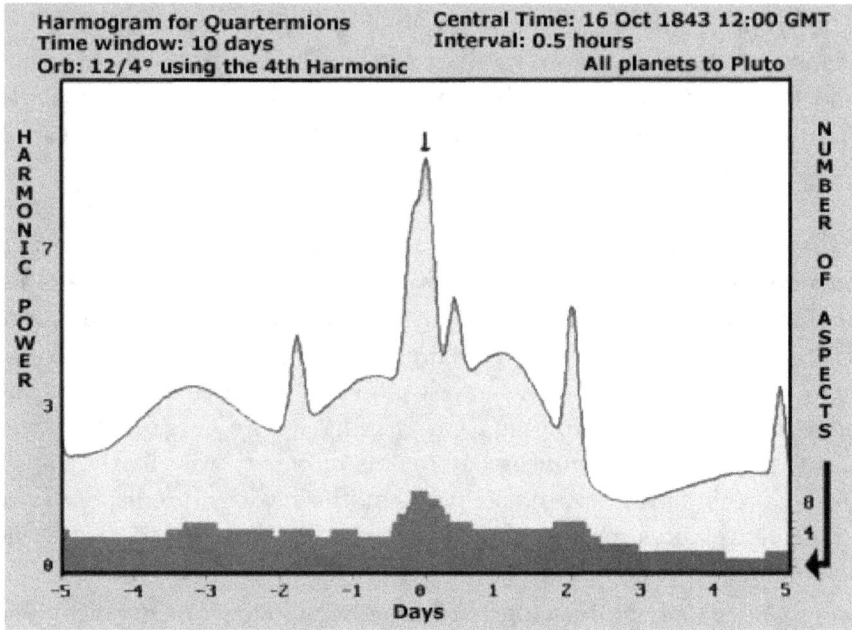

Figure 12.5: 4th harmonic harmogram For Quaternions ±5 days.

moments, for the reason that this writer is not capable of finding them, had been explained. He was interested by the harmograms, and asked pertinent questions about how far the pattern displayed was dependent upon the orb chosen for the harmonic. (In fact the peaks seem not to vary much with a greater or smaller angle taken as the base orb, but do vary somewhat with the central time chosen.) He then suggested that the Hamilton moment be examined, and in response to the question as to what harmonic number should be used, answered simply, 'four.' Thus the harmogram for quaternions came from a specific prediction made by B.C. He kindly ascertained the time at which the Royal Academy meetings at Dublin would have occurred on that day. It was his view that quaternions had been of much importance in the development of physics.

'It has been said that creativity requires one or more of the "three Bs" — the bath, the bed, or the bus,' wrote eureka psychologist, Howard Gruber.[4] Perhaps the bridge should be added to this list.

Hamilton was convinced that the operations of the mind,

Chapter 12. DNA and the Decile

including mathematical operations, are in perfect harmony with the operations of the physical world as God created it. To understand the mind is to understand the world, and vice versa,

explained his biographer.[5] What is disclosed by the fourth-harmonic harmogram points us to a deeper appreciation of this truth.

Pluto's Element

Now we look at a different kind of harmonic chart, which spans the passage of several years, based on the number five. We continue the discussion, begun in chapter 2, of the birth process of a new element. The chart for plutonium's coming-into-being helped focus ideas for the eureka study, as showing how a quintile in the natal chart of Seaborg related to another quintile in the plutonium chart.

We now re-examine that quintile between the planets Uranus and Pluto, which appeared so prominently in the plutonium chart, observing its formation as a process in time. The computer diagrams the meeting of these two distant spheres in the depths of space. We view this from a geocentric reference. Then, instead of mere uniform motion according to Kepler's second law, there is revealed a marvellous *pas de deux* of meeting and separation.

What is called a triple conjunction occurs when a planet becomes 'retrograde', reversing its apparent motion along the zodiac, so that it can meet or 'conjunct' a slower-moving planet thrice. It first overtakes, then crosses back, and again overtakes. Figure 12.6 shows a *fivefold* meeting, in which the first and fifth meetings merely touch without crossing. At the birth of this element which has so threatened life on earth, should we not expect some strong gesture from the womb of time?

The extent of Pluto's connection with plutonium has been discussed elsewhere.[†6] Here, we are concerned with its fivefold nature: it is abnormal for an atom to have five possible valence states, and

---

[†]For some deep ruminations on the theme, see Brian Taylor, 'The Discovery of Pluto: an Unbidden Omen' in *Orpheus, Voices in Contemporary Astrology*, Ed. Suzi Harvey, 2000, pp.247–330. Taylor perceived the essential connection between the Pluto-discovery and Plutonium-creation moments:

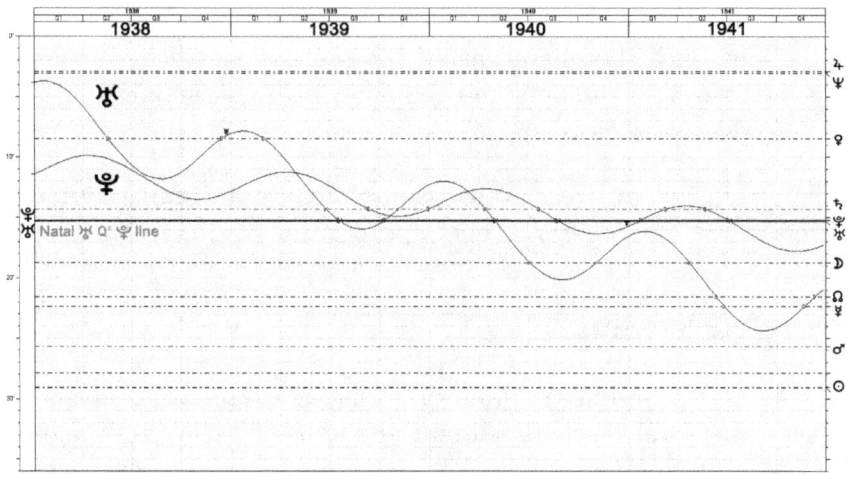

Figure 12.6: Uranus and Pluto quintiles, 1938–41, 10th harmonic (36°), and horizontal lines show Seaborg's natal chart.

five crystal conditions, but this ultra-heavy element does manage such, powerfully expressed in the circumstances of its coming into existence.‡7 The fateful new element was created on *the last of five quintiles.*

By dividing the circle of the zodiac into five equal sections each of 72° in length and overlaying them, we can make any quintile aspect show up as a conjunction. That is more or less what we will ask the computer to perform, but with one slight modification. Bisecting the Pluto-Uranus quintile aspect was Seaborg's natal Pluto, plus the transiting Moon — 'transiting Moon' means the Moon present at the event, in this case the creation chart, as opposed to 'natal', viz. that

Pluto-discovery	Ascendant	(Flagstaff, Arizona 1930)	3° 29' Leo
Plutonium-creation	Pluto	(Berkeley, California 1940)	4° 00' Leo
Plutonium-creation	Ascendant	(Berkeley, California 1940)	4° 35' Leo

This argues (as earlier with the Hiroshima chart) that the Pluto-discovery chart 'works'.
Cryptically, he noted that a straight line from Berkeley California to the Trinity test-site where the first (plutonium) atom-bomb was detonated in New Mexico passed through Flagstaff.

‡Plutonium 'undergoes no less than five phase transitions between room temperature and its melting-point' ... its ions are commonly in the 'III, IV, V & VI oxidation states, but also VII', i.e. it has five oxidation states, Seaborg, 1957, p.27 ref. 16.

Chapter 12. DNA and the Decile

present at the birth of the scientist. In order to display these also, we view the tenth harmonic pattern, dividing the zodiac circle up into ten equal 36° sections. We instruct the computer to plot solely the motion of Uranus and Pluto, excluding other planets.

The faster-moving of the two is Uranus. The five quintile meetings are evident. On the first and last they come very close: within 17' in 1938. The horizontal lines signify the natal chart of Glenn Seaborg, the solid black line being his close Uranus-Pluto biquintile. From the figure, one cannot tell whether the adjacent lines signify conjunctions, quintiles or deciles. (The natal chart has a Jupiter-Neptune biquintile and a Node-Mercury conjunction, though these have no significance here.) The small mark indicates the date of December 14th when the new element was formed, when both planets were retrograde.

The sequence of five quintiles moved from uranium fission by Otto Hahn in Germany in mid-1938 to the birth of the new element in early 1941 in Berkeley, California. In deep secrecy, both British and American scientists decided that the name of the new element should be plutonium, reaching that decision independently.[8] We should note that Pluto's appearance in 1930 was linked in the scientific sphere to more than just atomic fission: in 1932, radio astronomy and electron microscopy appeared on the scene, both concerned with perception of hidden orders of being, as in the same year the atom was opened up by Cockcroft and Walton, and the first particle of 'anti-matter' was then found, the positron.

This sequence of five quintiles has been linked to the political events of the day by the authors of *Mundane Astrology*: they noted of the first quintile, 'It is striking that Uranus was within 1° of orb of 72° to Pluto from June to September 1938, the period of German mobilisation (actually ordered 12th August) ... It was again within orb from April to June 1939 in the final build-up to World War Two'.[9] President Truman decided to press ahead with the Bomb on October 11th, 1939, and the middle of the five Uranus-Pluto quintiles fell three days later.[10]

The Galactic Centre on this decile diagram would be one degree to the left of the Pluto-Uranus biquintile line, next to the natal Saturn. The Galactic Centre in 1940 was 26° 1' of Sagittarius, i.e. opposite Seaborg's natal Pluto which was at 27° 14' Sagittarius.

Over the same months there was also occurring a triple encounter of Jupiter and Saturn which normally takes place once a century. They are known as Great Conjunctions. The first of these three chimed the previous summer, in August 1940, as Seaborg's team began their search for the new element. The second formed in October and the third in February, shortly after Seaborg's team had sent off their historic letter to the Physical Review reporting their finding of a new element. At the time of the discovery, these two spheres were going retrograde, as were Uranus and Pluto.

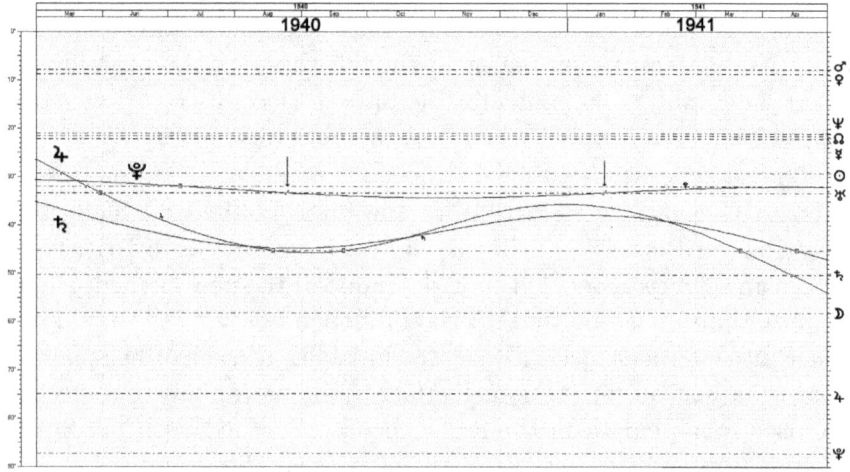

Figure 12.7: Jupiter-Saturn in square To Pluto, 1940–41, 4th Harmonic (90°).

The triple conjunction moved into a square aspect with the new planet Pluto. To investigate this, we instruct the computer to use a fourth harmonic scale, which divides the zodiac into four 90 sectors and superimposes them (figure 12.7). Thereby is revealed how plutonium emerged at the moment when Pluto approached closest to a square aspect with the Great Conjunction. An opposition of Pluto to the natal Uranus is indicated by a thin vertical arrow.

The plutonium creation was a genesis-moment of a rare kind,[§] in that it had a precise timing at Berkeley, California of 20:00hrs PST 14th Dec 1940. As we have looked at the unfolding of some very slow aspects within the plutonium-creation chart, let us now consider the remarkable fine adjustment it has within minutes. There were two

---

[§]14th Dec 1940 8.00 p.m. PST California. See http://www.skyscript.co.uk/metal9.html.

Chapter 12. DNA and the Decile

notable features of the chart in this respect. One was the position of the new planet Pluto conjunct the ascendant within 26′, and the other was the position of its Moon on Seaborg's natal Pluto within 5′. Each of these events defines the timing of the chart to within a minute. These are different timescales over which a birth process can be viewed: the slow meeting of the outer planets takes years, while a planet rising over the ascendant takes minutes.

In Glenn Seaborg's chart, Pluto appears as on the IC, the lowest part of the chart, and in a close square to the ascendant. We only have Seaborg's time of birth to within the hour (around 04:00hrs local time), so one cannot be more precise. Perhaps he was born a few minutes earlier, which would give a more exact positioning of Pluto as conjunct the IC and square the ascendant. At any rate, the then-undiscovered planet was lurking at the base of his chart in a forceful aspect. We have already observed how it stood in a biquintile to Uranus, within minutes. (The angle was 143° 56′, a biquintile being 144°.) For most charts we lack the marvellous time-definition which this pair of event and natal chart offers us. We see between them the Plutonic linkage. Figure 12.8 shows the 5th harmonic synchronies: the natal plutonium creation biquintile, with his Uranus conjunct the descendant of the plutonium chart, moving to a quintile at the creation-moment with Pluto conjunct the ascendant, such that it has moved a decile aspect (36°) since Seaborg's birth.

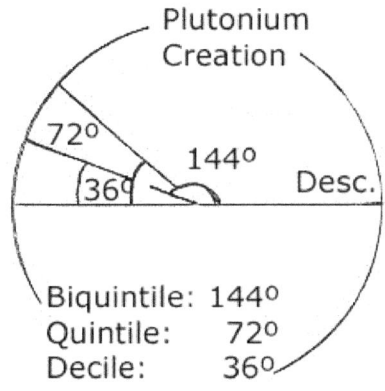

Figure 12.8: Plutonium creation and quintiles.

Natal Positions
Uranus 3° Aquarius
Pluto 27° Gemini

Event
Ascendant 4° Leo
Moon 27° Gemini
Pluto 4° Leo
Uranus 23° Taurus

Seaborg wrote books about plutonium, the first synthetic element

275

to be seen by man. He became chairman of the U.S. Atomic Energy Commission and advisor to President Kennedy. It is in accord with the theory of astrology that his natal chart should show so clear and precise a link with that solemn and perilous moment, when the formation of the new element was commenced.

Seven, Ten and DNA

A 25 year-old American, James Watson, was assembling some cardboard cut-outs in a molecular biology laboratory in Cambridge, in the year 1953. Having managed to obtain the latest X-ray photographs of a purified substance, he was mulling over the curious properties of hydrogen bonds, which were weak and easily broken, wondering how four different kind of amino-acid molecules could be fitted together in pairs. Finally, a shape never dreamed of hitherto dawned upon him, for the huge double helix molecule DNA, the molecule of life.

After months of struggling with the enigma, it suddenly clicked. He realised, at around 11 a.m. on a Saturday morning, that each base-pair, each link-up of the protein code, was rotated at a 36° angle to the previous one, so that if one looked down the length of the giant molecule, the helix revolved once every ten steps.

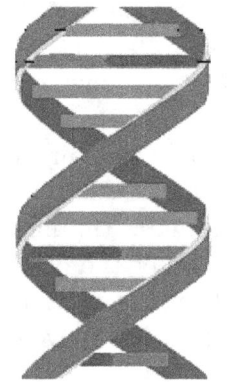

Figure 12.9: DNA double helix.

The double helix of DNA entered a human mind on that morning. A molecule with a hitherto unheard-of structure was then envisioned. The world's experts on the subject were pipped at the post by a 25-year old American, James Watson. Cambridge again became the epicentre of the scientific world, as it had been some decades earlier when the atom was split there. Early histories did not locate the date of the incident, and it was only found by Judson in *The Eighth Day of Creation* (1983) as having been Saturday, February 28th of 1953.

The author, as a student at the college of Corpus Christi at Cambridge a dozen or so years later, was accustomed to frequent the Eagle tavern adjacent to it, where Francis Crick was supposed to have emerged that climactic Saturday morn-

Chapter 12. DNA and the Decile

ing to announce the great discovery — surely, rivalling Archimedes' moment.[11] The pattern of the heavens at that moment helped a specific idea to form in the mind of one prepared. The next chapter compares the charts of the two collaborators.

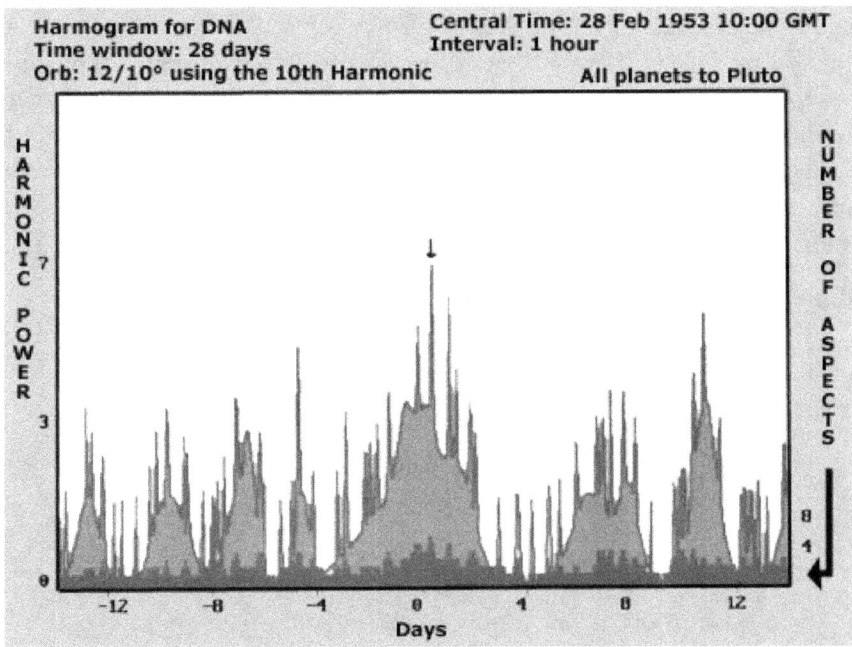

Figure 12.10: DNA 10th harmonic Harmogram ±2 weeks.

We should expect that the heavens above would reflect this tenfold structure comprised of 36° angles, at that moment in time. We are here somewhat reversing the ancient Hermetic maxim, 'As above, so below'. Two distinct harmonics peaked at that historic moment, the tenth harmonic being the stronger. Ten is the number specific to what was discovered at that moment in time, as the helix of DNA coils round once per ten steps. Each step constitutes a base-pair, a pair of nucleic acids which contain the 'genetic code.' The so-called base-pairs are held together by hydrogen bonds, which cleverly open up when the time comes for cell division, so that a copy of its information can be made. The numbers of the genetic code, such as four for the types of nucleic acid, were not discovered in this historic moment. What then materialised was a three-dimensional structure,

EUREKA

of a tenfold symmetry.

The harmogram (figure 12.10) covers a period of one month, showing that the tenth harmonic was peaking at a unique moment in time. The number seven, having a more general significance for times of scientific inspiration, also peaks at this time. The septile harmogram (figure 12.12) shows how well the timing of the E-moment is defined by the septile peak. At the base of the harmogram, the count of septile aspects shows no peak at the required moment on the Saturday morning. The line showing what we call 'septile power' responds to the closeness of orb, in addition to the overall number of aspects, and so gives a more sensitive indicator of the inspiration-quality of the time.

Table 12.1: The seven 10th-harmonic aspects at the DNA E-moment.

No	Planet	Aspect	Planet	Orb				
1	Moon	decile	Mars	15'				
2	Mercury	tridecile	Uranus	50'	where:			
3	Mercury	biquintile	Pluto	10'	decile	= 36°	(1 × 36°)
4	Venus	opposition	Neptune	50'	quintile	= 72°	(2 × 36°)
5	Uranus	decile	Pluto	1°	tridecile	= 108°	(3 × 36°)
6	Venus	quintile	Node	20'	biquintile	= 144°	(4 × 36°)
7	Neptune	tridecile	Node	30'	opposition	= 180°	(5 × 36°)

Our estimation of the timing of the event as 10 a.m. in the morning came from the account given in 'The Double Helix' by James Watson. He describes how he arrived in the morning and set out the base-pairs, and how after it all came together, Francis Crick turned up, and they spent a while sorting it out.

Then at lunch-time Francis Crick 'winged his way into the Eagle to tell anyone within earshot that we had found the secret of life.' Crick's version of events[12] contained (rather unsportingly) no memory of such a thing, but we used it nonetheless as a basis for placing the event at between 10 and 11 a.m.

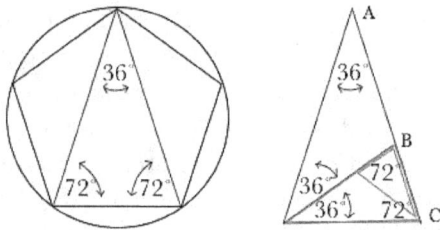

Figure 12.11: The pentagon and the Golden Triangle.

278

Chapter 12. DNA and the Decile

These seven 'tenth-harmonic' aspects are the defining feature of this supreme Eureka-moment. Astrologers associate the lunar nodes with fate and destiny, as is here highly relevant.

Astrologers would generally ignore a single decile aspect as being too weak: such aspects were relegated by Kepler to almost the lowest level of power, below the quintile and above only the octile.[13] But, as the pentagram is more symmetrical than the octagon, so he found the decile to be more effective than the octagon. Structurally, the decile and tridecile aspects, as both appeared twice at the DNA E-moment, form the inner angles of the pentagram and pentagon. The pentagram is a regular geometric figure that is full of Golden Ratios, for example its sides intersect each other in that ratio.

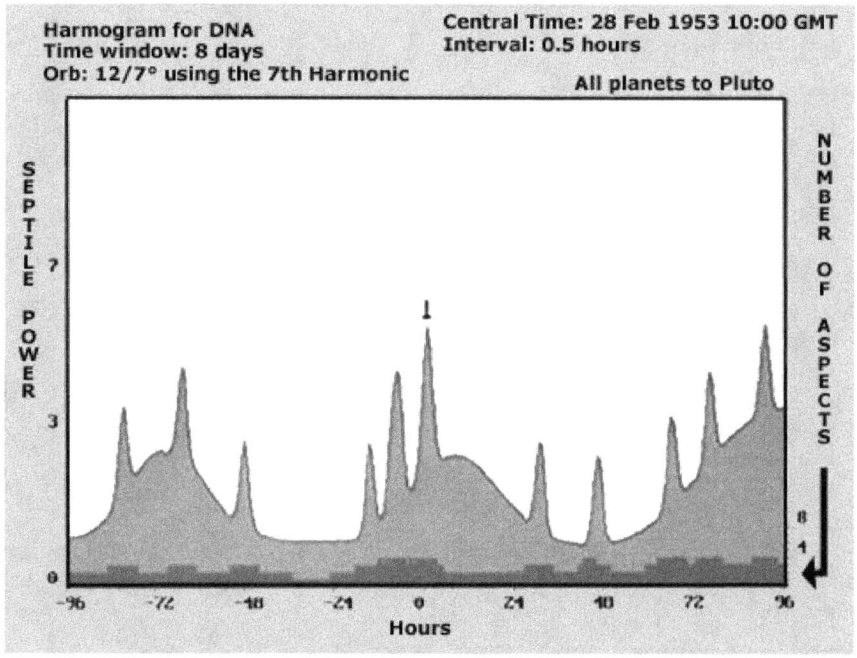

Figure 12.12: DNA 7th harmonic ±4 days.

The Golden Ratio is expressive of the unfolding of living things, as in the spiral shapes in flowers and the rotation of their leaves up the stem; we may hardly view this as unconnected with the ten-step rotation within the coil of DNA, the molecule of life.

We may picture the pentagram/pentagon as woven of five 'golden triangles.' This quite special triangle has a decile at its top, and two quintile angles at the base. So this is quite a remarkable quality of the quintile, of forming the Golden Ratio. It may help us to apprehend the 'creative' nature of the quintile aspect. By bisecting one of its base-angles to make two deciles, one begins to weave a pentagram.[14] It may help to think of the decile angle as being roughly, within half a degree, the small angle in the 3-4-5 triangle.¶

There is one other E-moment for which two distinct harmonics converged, and that, as we have seen, is the moment of rapture when Kepler realised the third of his planetary laws, when the third and seventh harmonics then converged.

References

1. D. Cochrane, *Astrology for the 21st Century* FL 2002, p.103.

2. This correspondence of 1608 is described in: Judith Field, 'A Lutheran Astrologer: Johannes Kepler', *AHES*, (1984), p.202.

3. Thomas L. Hankins, *Sir William Rowan Hamilton*, 1980, p.136.

4. H. E. Gruber, 'On the Relation between Aha Experiences and the Construction of Ideas.' *History of Science*, 1981, xix, p.42.

5. Hankins (ref. 3), p.30.

6. N.K., 'Pluto and Plutonium', *The Astrological Journal*, Autumn 1984, http://www.skyscript.co.uk/metal9.html.

7. *Chemistry of the Actinide Elements*, J. Katz & G. Seaborg, 1957, p.265.

8. R. W. Clarke, *The Greatest Power on Earth*, 1980, p.114. It remained unnamed for two years, until March 19th, 1942, when the name 'plutonium' was suggested.

9. Campion, Baigent and Harvey, *Mundane Astrology*.

10. R. Rhodes, *The Making of the Atomic Bomb* 1986, p.313.

¶This approximation is used by D. Bunker, in *Quintiles and Trideciles, the Geometry of the Goddess* (Penn. US 1989) arguing that the quintile and tridecile aspects are 'the ultimate creative aspects in the chart' (p.47). She reached the insight on which her book was based in November of 1987, the same year as the present Eureka project was initiated (note 9, Ch. 2) and, commendably, gave a chart for her E-moment.

References

11. Only after Francis Crick's death at 88 years of age, did a dubious story start to be heard, that he had discovered the structure of DNA while high on LSD. Crick does seem to have commented on this: his comment was, 'Print a word of it and I'll sue.' One is bound to add, that the year 1953 was at least a decade before the idea of having LSD 'trips' arose. http://www.hallucinogens.com/lsd/francis-crick.html.

12. J. Crick, *What Mad Pursuit*, 1988, p.77.

13. Kepler, *Harmonice Mundi* 1618, 1997 p.354.

14. M. Ghyka, '*The Geometry of Art and Life*' Dover NY 1997, p.24.

CHAPTER 13

The Synastry of Invention

SYNASTRY IS AN ASTROLOGICAL TERM referring to the linkage between two charts. It should be able to depict the relationship between persons involved in a collaboration, or making the same discovery. It involves comparing two natal charts, as opposed to the transits to a natal chart reviewed in chapter 11. If unity of purpose existed between the collaborators, we might expect conjunctions between the natal chart planets, or if bitter strife then an opposition, or for the harmonious flow of shared energy, a trine. Thereby we may hope to gain information about the chemistry of a relationship, even without having the times of birth of the persons concerned.

For example, there was a famous controversy over the discovery of calculus, as to whether the British fluxions or the German differentials came first. Experts now view them as separate and independent inventions, but at the time fierce debate raged between British and continental mathematicians. The two sages involved — Sir Isaac Newton, President of the Royal Society, and Baron Gottfried von Leibniz, President of the Berlin Academy of Sciences — had Mercury opposition Mercury (2°), plus Sun opposition Sun, which is a synastry of conflict by any standards. But also they had Jupiter trine Jupiter, indicating perhaps that they could not help coming up with the same ideas.

Mars against Mars

One would expect the sparks to fly in a synastry based on Mars-contacts. A case in point is the relation between Tycho Brahe and Kepler, which was indeed stormy. Koestler described their first meeting:

> At last, on February 4 1600, Tycho de Brahe and Johannes Keplerus, co-founders of a new universe, met face to face, silver nose to scabby cheek. Tycho was fifty-three, Kepler twenty-nine. Tycho was an aristocrat, Kepler a plebeian; Tycho a Croesus, Kepler a church-mouse; Tycho a Great Dane, Kepler a mangy mongrel. They were opposites in every respect but one: the irritable, choleric disposition which they shared. The result was constant friction, flaring into heated quarrels, followed by half-hearted reconciliations.[1]

It is appropriate to find a Mars opposition Mars (2°) between the charts of these two. In their tempestuous relationship we see one aspect of the great Mars-endeavour on which the two collaborated. From studying Mars' orbit, Kepler after years of struggle discovered elliptical orbits. The hardest bit was probably getting the data off Tycho Brahe in the first place. Things exploded in a bitter argument on April 6th, 1600, a few months after Kepler had arrived, of which his biographer Caspar said: 'Kepler, who liked to compare his irritability to that of a mad dog, had indeed behaved like one ...'[2]

On the date of that row, Mars formed a sextile aspect to Kepler's natal Mars. Tycho Brahe demanded a written apology — and he got it. Perhaps Kepler noticed this aspect, for later on he remarked, 'Regard this as certain, Mars never crosses my path without involving me in disputes and putting me myself in a quarrelsome mood.'[3]

On Kepler's arrival, Mars had just passed its opposition to the Sun — when it comes nearest to the Earth — and Brahe had prepared a table of its oppositions since 1580. A month or so after Kepler's arrival at Benatek castle, he persuaded Brahe to allow him to tackle the orbit of Mars. Brahe was reluctant to let the newcomer see his hoard of data. Kepler boasted that he would be able to solve the orbit in a short time, and whether or not he believed this, it was effective in persuading the moody Dane to part with his data. It was the only planet with a large enough eccentricity for an elliptical orbit to be detectable: any other, for example Venus, and there would have been no hope of making the discovery which banished circles from the heavens. Traditionally, Mars had been called 'the unobservable star' because of its impossibly irregular motion. Here is how the modern astronomer Sir Fred Hoyle expressed his admiration for Kepler's achievement:

> Kepler fitted the task of advancing Copernican theory, not only in stature, but also in the demonic energy with which he tackled the problem of Mars. The sheer volume of calculation which he carried through fits him to be called the iron man of science.[4]

Kepler and Newton

Newton's theory of universal gravitation was founded upon Kepler's three laws. From them he deduced his new scheme of the universe. Newton lived through one Uranus cycle and Kepler through two Saturn-cycles. Kepler's natal Saturn was conjunct Newton's natal Uranus, so that these two cycles returned at death to the same birth positions, pointing to a deep connection in their destinies. Thus, their synastry is shown in table 13.1.

Table 13.1: Kepler and Newton synastry.

	Kepler		Newton	
Birth:	Saturn	13° 30′ Scorpio	Uranus	12° 00′ Scorpio
Death:	Saturn	10° 00′ Scorpio	Uranus	13° 30′ Scorpio

In addition, they were both born over Christmas so that their natal Suns were conjunct, a mere degree and a half apart. Overall, there appears 'a truly remarkable similarity.'[5]

The Uranus cycle marked out major stages of Newton's creative endeavours, as we have seen. Kepler's Saturn return at the midpoint of his life fell on 9th November 1601. The previous year he arrived in Prague from Linz and commenced working for Tycho Brahe. The latter died on October 21st, 1601, and then on 6th November, Kepler acquired the post of Imperial Mathematician for the Holy Roman Empire. Following Brahe's death, he was able to acquire the priceless astronomical data that Brahe had amassed from his time in Denmark. Also, following Brahe's death, in the next couple of months, he composed his *De Fundamentis Astrologiae Certioribus*, translated as 'On the More Certain Fundamentals of Astrology.' These are all appropriate events for a Saturn return.

Mercurial Strife

We have claimed that the opposition between the Mercury positions in the charts of Newton and Leibniz was relevant to the disputes which took place between them. Our authority for such a viewpoint surely has to be John Aubrey: In the sole vestige of astrology remaining in the modern version of his classic *Brief Lives*, we learn:

> And the controversies that raged over *Arithmeticall Problemes* reached such a pitch of emotion (particularly when Hobbes thought that he had squared the circle and Dr Wallis knew that he had not) that poor Aubrey was driven to the conclusion: *sure their Mercuries are in* □ *or opposition.*[6]

With a comment like that, it seems worth while to examine the relation between the two persons Aubrey has specified, the philosopher Thomas Hobbes and the mathematician Dr John Wallis (see table 13.2). These arguments took place around the 1670s, an exciting time for mathematicians who were finally discovering ways of 'squaring of the circle' or as we would nowadays say, finding the value of π, something that had baffled mathematicians down through the ages. John Wallis in his book on 'Algebra' described some of these methods, as being an achievement of the English mathematicians. He was the professor of Mathematics at Oxford University. (He held what was called the Savilian chair of mathematics. Sir Henry Saville, who founded it a few decades earlier, specified that anyone holding it was forbidden to teach astrology. In the old days, before the Civil War, mathematics students would normally learn astrology.)

Wallis was a fairly abrasive character, who fell out with most of the continental mathematicians through the claims he made for

English mathematics. Yet, he was said to possess a 'rugged charm' which made up for it. Thomas Hobbes was one of the best-known figures of the age and regarded as a great wit and conversationalist. He was not invited to join the Royal Society, Aubrey explained, because of one or two enemies (such as Wallis) within the institution. Wallis, who was religious and became the King's chaplain after retiring from his chair at Oxford, did not want the Society to be seen as supporting Hobbes' materialistic philosophy, which was becoming notorious.

Was Aubrey right? His comments suggest a challenging Mercurial synastry between them which was indeed the case. However, their Mercuries stood *not* in square or opposition, but in a trine! Now a trine is supposed to promote concordance, not bitter strife. But their two Mars positions were also in trine, and this trine was in trine to the Mercury trine! For their synastry, the computer gives us:

Table 13.2: Mathematical dispute of Wallis-Hobbes synastry.

Hobbes	Aspect	Wallis	Orb
Mercury	trine	Mercury	2° 00′
Mercury	trine	Mars	4° 27′
Mars	trine	Mercury	3° 26′
Mars	trine	Mars	3° 01′

This seems a fine celestial background for mathematical arguments, no doubt blending quicksilver logic with Martial passion. In Restoration England, mathematics was a subject of keen public interest and debate. In addition, the two had a Uranus trine Uranus (1°) perhaps relevant to a shared interest in recent discoveries.

The Discovery of Neptune

The diary of John Couch Adams records the date when he made the key decision of his life, to seek for the position in the sky of what was rumoured to be a new planet:

> 1841. July 3. Formed a design, in the beginning of this week, of investigating as soon as possible ... the irreg-

ularities in the motion of Uranus ... [in relation to] an undiscovered planet beyond it ...

In Adams' natal chart, Neptune was then receiving a transit of the planet Saturn within a degree, i.e. Saturn was passing over the position of Neptune at Adams' birth, which is relevant for a scientific study of the subject. His birth in 1819 was close to the Uranus-Neptune conjunction, and the mathematical computations he had to perform were centred on that event. Uranus wasn't where it ought to be, it refused to fit onto a Keplerian-ellipse orbit, staying an arc-minute or two away. The perturbation was *caused by* that conjunction of 1820, with Neptune pulling on Uranus during the decades centred on that date. Adams had that key conjunction within his own birth chart, and Saturn was passing over that zone of the heavens when he decided to tackle the problem. At 22 years of age, he was then experiencing his Uranus-square return, within about one degree.

By September 1845, he had reached some sort of solution for its position, in the constellation of Aquarius. However, he was remarkably backward about coming forward and documentary evidence that he had left a note on the subject, with the Astronomer Royal or anyone else, remains dubious.[7] There are one or two undated and unaddressed bits of paper from him alleged to belong to this period, but no letter to anyone about a discovery — he wrote no letter to anyone about his prediction until September 1846.

In July of 1846, a public prediction of the new planet's position was made by the distinguished French astronomer Urbain Leverrier, which led to a clandestine British search, using a high-powered observatory at Cambridge. In this search, they were encouraged by some indication which the young Adams had also given them. On July 9th, the Astronomer Royal George Airy wrote to the Rev. James Challis, the astronomer who was about to start the sky-search:

> You know I attach importance to the examination of that part of the heavens in which there is a possible shadow of reason for suspecting the existence of a planet exterior to Uranus.

After the discovery, he claimed to have had a somewhat stronger degree of confidence over its existence and position. The Cambridge

Chapter 13. The Synastry of Invention

Observatory had been plodding on for six weeks, and then the Berlin Observatory found it in half a hour, on the night of September 23rd. It was an awesome British failure. During the week after the discovery, reports of the British case for co-discovery started to emerge. Mr Adams' case would shortly be published. Explanations for his delay harped upon his modesty and reticence. In the second week in October the storm broke, as the French responded to the retrospectively-constructed British claim to have co-discovered the new planet. Still no-one declared what Adams' predictions of the new planet were supposed to have been.

The British case for co-prediction was finally made at a Royal Astronomical Society meeting of November 13th — six weeks after its discovery! Neptune became stationary (i.e. appeared not to move in the sky) after the first week in November, mere days before that solemn meeting, when the astronomers Airy and Challis and the mathematician Adams presented their case. The marble bust of John Couch Adams resides next to that of Isaac Newton in Westminster Abbey — yet, when all is said and done, Britain's case for having an independent prediction of Neptune remains a fishy tale. In the 1960s, Britain's 'Neptune file' vanished, containing all the vital documents, from the Royal Greenwich Observatory's archives at Herstmonceaux, then finally turned up again in 1999, in Chile (the Chief Assistant to Britain's Astronomer Royal had made off with them). Astrologers are well-placed to apprehend the pall of Neptunian miasma which hovers over the British story.

Adams' chart was loaded with Neptune aspects. His outer planets were locked together in a single square: Neptune conjunct Uranus (5°) was square Saturn and conjunct Pluto (2°), with Jupiter close to being at the midpoint of the square, and Mars in exact trine to his Neptune (48'). His chart was oversaturated with Neptune aspects as worked against him, preventing anything from happening. The usual story is that British astronomers viewed him as a dreamer and never took him seriously, while in reality his views were uncertain and changeable. The Neptune-Uranus conjunction at his birth in 1819 (exact in 1820) was the last prior to that of 1992.

Neptune's position in the natal chart of Adams was conjunct the Saturn of Leverrier's chart within 2°, and the Sun of Adams was opposite Leverrier's Neptune within 3°, as well as being conjunct his ascendant (4°). That is a profoundly appropriate synastry. Adams' Uranus was conjunct Leverrier's Saturn, as well as being conjunct Leverrier's descendant, plus his Jupiter was conjunct Leverrier's MC, giving a remarkable tie-up between the charts of these two people who had never met, to whom the same idea occurred. Leverrier's chart is timed while in the case of Adams' chart only the birth date is available.

Neptune was discovered as it wove its triple conjunction with Saturn, on the middle of the three conjunctions. It then appeared at once, leaping to the eye at the Berlin Observatory. This aspect turns up thrice in our story: as a transit when Adams made his big decision in 1841, then in the sky at the discovery, and finally in synastry between the two co-predictors. Traditional accounts have difficulty in explaining how no-one saw it earlier, and how Challis managed to spend six weeks not finding it, when a prediction of its position to one-degree accuracy had been published. It needed the fixity of being in conjunction with Saturn to enable that sea-blue sphere to be seen.

This discovery has features comparable to the invention of anaesthesia, which happened within weeks of Neptune's discovery on the other side of the world. In both cases, the Saturn-Neptune conjunction was the trigger for the discovery. Morton who discovered anaesthesia was born within a few months of Adams, having a similar powerful constellation of Neptune aspects, and under different circumstances also failed to gain credit for his discovery.

Neptune's discovery consolidated the theory of gravity whereby it was predicted, so we may regard the conjunction with Saturn as appropriate.

Darwin's 'Origin'

> *'Without the stimulus of Darwin, there might have been no Wallace, just as without the stimulus of Wallace, Darwin might never have got around to formal publication.'*
>
> Loren Eiseley.

We here review the four dates:

Chapter 13. The Synastry of Invention

Darwin's eureka insight	Sep 28, 1838	Uranus at 9° 31′ Scorpio
Wallace's insight	Feb, 1858	
Darwin receives Wallace's letter	Jun 12, 1858	Uranus at 5° 52′ Gemini
Publication of Darwin's *Origin*	Nov 24, 1859	

in terms of their Uranus transits.

The vision of evolution by natural selection dawned upon the twenty-eight year old Charles Darwin on September 28th, 1838, with Uranus conjunct his natal Mercury (46′) and trining his natal Uranus within ten minutes of arc. This means that he had the idea within a day or so of his Uranus-trine return, as would have activated the Uranus-Mercury trine in his natal chart. We saw how there were five septile aspects in the sky, an appropriate day for a brainwave. For the next twenty years he kept quiet about his views, amassing copious notes on the subject. There things would have remained, but for an intervention by the naturalist Alfred Russell Wallace, then in the depths of the Malay jungle.

Much the same notion dawned upon Wallace, in February of 1858, with Uranus at $25\frac{1}{2}$° Taurus, conjunct his natal Jupiter to one degree — and a degree from the fixed star Algol then at 24° Taurus (Algol has traditionally enjoyed the reputation of being the most evil star in the firmament, forming the constellation image of the Gorgon's head). Wallace was swinging on his hammock in a fever near to death on a tropical island — or, so he claimed (see chapter 6: The Moment Of Illumination) — when he recalled the pessimistic work by Malthus he had read years ago, on how human population must always multiply to excess:

> Vaguely thinking over the enormous and constant destruction this implied, it occurred to me to ask the question, 'Why do some die, and some live? and the answer was clearly that on the whole the best fitted live ... Then it suddenly flashed upon me that this self-acting process would improve the race ... the fittest would survive. Then at once I seemed to see the whole effect of this.[8]

He succumbed to the nightmare illusion, that all Nature's varied forms had derived from a blind and random process.

On June 12th, 1858, Darwin received the shock of his life with a letter from Wallace. He claimed to discern the very headings of his notebook in that letter. 'I never saw a more striking coincidence' was his comment. Later that year, the theory of evolution by natural selection was proposed at the Linnaean Society by the two of them, Darwin and Wallace. The trauma of Wallace's letter set events in motion for publishing the *Origin of Species* the next year.

Darwin received the letter after he had been mulling over the general theme for twenty years. The two naturalists had previously corresponded about what was producing the diversity amongst species, sensing that they were in tune in a general way, but on that day it became apparent that the most original ideas in their two lives, were one and the same. If we look at the synastry between the natal charts of these two in relation to June 12th, 1858, then Uranus was in conjunction both to Mercury and to Jupiter, and was in a 'Promethean' mode of behaviour, as Richard Tarnas has described it. Mercury was in one degree of Wallace's Jupiter, Jupiter was in one degree of opposition to Darwin's Saturn, and Uranus was in opposition to Wallace's Moon to within a fraction of a degree. These were powerful transits for a key moment of his life:

> A man pursuing birds of paradise in a remote jungle did not know that he had forced the world's most reluctant author to disgorge his hoarded volume.[9]

The Origin of Species was published on the 24 November, 1859, selling out on that day then speeding through six further editions. On its publication date, Uranus opposed a new Moon, i.e. Sun conjunct Moon, which became exact at 1 p.m. that day — at 2° Sagittarius, a mere degree from Wallace's Moon and Darwin's Saturn. Uranus at 6° of Gemini was exactly opposing Darwin's natal Neptune, both being aligned with the big star-axis of Antares-Aldebaran. Just as in February 1858, Uranus was precisely conjunct Wallace's Jupiter, so now as Darwin became the central figure, Uranus moved into exact opposition to his Neptune-Saturn conjunction.

Persons sceptical of the Darwinian theory may view the *Origin's* Neptune square Mercury (1°) and exact trine to Jupiter (38′) as pertinent to the illusory nature of what then appeared, as Mr Elwell indicates in *The Cosmic Loom*.[10] Did Neptune conjunct Saturn in Darwin's chart help in dreaming up the theory? As Darwin's *Origin of Species* confessed: 'Why then is not every geological formation and every stratum full of such intermediate links [between species]? Geology assuredly does not reveal any such finely graduated organic chain; and this perhaps is the most obvious and gravest objection which can be urged against my theory.' It was indeed. The absence of the assumed intermediate forms in the fossil record was *the* problem — the 'absolute rebellion of the facts', as Bishop Wilberforce in his review of the *Origin* described the situation.

The Stars of Conflict

After publication, the spotlight moved onto Thomas Henry Huxley, as Darwin was too ill to take part in any debates. Huxley had a clear idea of what was involved, as his Saturn-Moon opposition, at the crux of a beautiful 'kite'- formation in his chart, was lined up with Darwin's Neptune and the *Origin's* Uranus: ($Moon_H 7\frac{1}{2}°$ Sagittarius, $Saturn_H 7\frac{1}{2}°$ Gemini, $Neptune_D$ 7° Sagittarius) — and the big star-axis Antares-Aldebaran! Huxley's chart is timed, and has its Moon conjunct Darwin's Neptune to 49′, with Saturn opposite that Neptune to 57′, and Venus opposite that Neptune to 50′. That is a powerful synchrony for two persons advancing the same world-conception. It was made even more so by being aligned with the two first-magnitude red stars Antares ('heart of the scorpion') and Aldebaran ('the bull's eye'), which are in the zodiac and precisely opposite each other in the sky. They stood at seven and a half degrees of Gemini-Sagittarius. As this star-axis has been associated with war and strife,[11] so Huxley became the storm-centre of the tremendous battles in Victorian England between theology and what was then becoming called, 'science.' Jupiter rising assisted his triumph in such debates. The opposition between the Saturns of Darwin and Huxley was rather wide at $4\frac{1}{2}°$, more than astrologers usually allow for synastry links, yet may be worth noting.

Huxley's Venus was closely conjunct his Saturn at 6° Gemini, so he experienced a powerful Uranus transit as the *Origin* appeared (see figure 13.1), its appearance being an event of supreme importance in his life. His Mercury was 27° Taurus, a mere half-degree from Wallace's Jupiter. 'I am sharpening my beak and claws in readiness', he wrote to Darwin. He became known as 'Darwin's bulldog'.

EUREKA

Strange forms were emerging from the fossil record, challenging the theological scheme of things. In Victorian England, Huxley became renowned for the clash of controversy which he brought into public debate. It was said, 'Where there was strife, there was Huxley.' He had a Sun conjunct Mars (6°), plus a close Mars trine Uranus (36'). He is remembered for pungent epigrams such as 'Extinguished theologians lie about the cradle of every science as the strangled snakes beside that of Hercules;' and found an inherent warfare between science and theology. A notable opponent of his in debate was Bishop Wilberforce ('soapy Sam'), who died while out horse-riding, being thrown off his horse and dashing his brains out. When Huxley heard the news, his typically caustic comment was: 'His brains finally made contact with reality, and the result was fatal.' Contemporaries noted Huxley's daemonic energy, as he pulled down the supernatural pillars supporting Victorian morality.

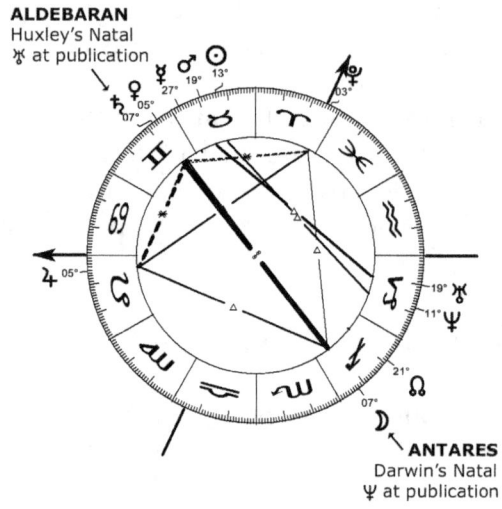

Figure 13.1: Natal Chart of T. H. Huxley, 4 May 1825 NS 09:30hrs LMT Ealing, UK (5° orbs used.)

Huxley was offered the Presidency of the august British Association for the Advancement of Science in August 1869, the month in which his Uranus opposition became exact. He was then 44 years of age. Huge crowds were turning up for his lectures, with newspapers alluding to him as the 'Prime Minister of Science', and the spectator dubbed him 'Pope Huxley' due to his air of infallibility. It was then that his 'Lay Sermons' were published, being viewed as the finest British essays on science since those of Lord Bacon. Also, in that month of August, the centre of his Uranus return, he first coined the term 'agnostic,' as his own label. Huxley's life thereby fits the Tarnas model about the Uranus-opposition return, as we analysed it

Chapter 13. The Synastry of Invention

in chapter 11: Destiny and the Uranus Cycle.

Huxley was, as his modern biographer concluded, 'never at peace unless he was fighting.'[12] Having Antares-Aldebaran as the central axis of his chart gave his life a continual supply of high-voltage electricity, as could not have been easy to live with. It has been said of these stars, 'What does tend to show up when these two stars are both engaged on the horoscope of a conflict ... is that peace rarely ensues before the total and crushing defeat of one side or the other. And it is a war that escalates to huge proportions, even to a world scale.'[13] The strife that Huxley engendered was never personal, did not happen in his family, but was with social injustice and the whole order of things.

Huxley was the most influential scientist that there has ever been in public debate. The word 'scientist' was first applied to him, in the year 1876, and in the next decade came to be applied to lesser mortals — relevant to the focal position of Saturn in his chart. His life is epitomised by a comment of Richard Tarnas, a propos of the birth charts of eminent and revolutionary characters:

> ... the presence of a Mars-Uranus aspect seemed to signify an emphatic joining together of the Mars impulse of aggression, militancy, and assertiveness with the Promethean tendency towards radical change, revolution, and liberation.[14]

Gnosis and *scientia* are synonyms, as the Greek and Latin words for knowledge. One cannot help reflecting upon these two terms, first applied to Huxley and now used by all the world. He was a 'scientist', i.e. one who claimed to know, to have knowledge, but an 'a-gnostic', i.e. one who did *not* have gnosis, knowledge — in the Greek sense! In the latter sense, knowledge meant an inner or personal experience, an accord between one's own being and the scheme of things — that which ever since Bacon and Descartes has been excluded from whatever 'science' means. Perhaps the present treatise is more 'gnostic' than 'scientific', though that could be taken to imply a denigration of its value. The Greek word has come to mean something which cannot be properly shared, whereas science is public knowledge. We may return to this issue.

'Lucky Jim' finds DNA

James Watson and Francis Crick collaborated to establish the structure of DNA, with the synastry between their charts being activated by Uranus, over the couple of months when they immortalised their names. They worked well together but soon split up with rows after the discovery. Crick was livid over Watson's racy and egotistical account, *The Double Helix*, and consulted a solicitor over whether to sue for libel.

Crick had a Venus-Saturn conjunction which lay across Watson's Pluto-descendant conjunction (Crick has no idea of his birth time, but Watson's is known to the minute, enabling us to position his ascendant.* The relevant degrees in the sign of Cancer were:

14° 39'	Uranus	of DNA February (28th, 1953)
15° 47'	Saturn	of Crick
15° 00'	Pluto	of Watson
14° 05'	Descendant	of Watson

Crick's Venus stood at 19° of Cancer. Uranus at the discovery was retrograde, becoming stationary a couple of weeks later. A Uranus transit of Saturn and Pluto does seem a good start for discovering the 'secret of life.' At the end of January, as the chase for the wonder-molecule hotted up, Uranus was transiting Crick's Saturn, then moved towards Watson's descendant, without quite reaching it. It turned at a mere ten minutes away from the position of the horizon when James Watson was born, and commenced going direct. On the 25th April when the famous article appeared in *Nature* describing the breakthrough, Uranus had reached conjunction again with Watson's Pluto. One cannot but be impressed by the precision of these transits.

They called him 'Lucky Jim', for he had a knack of being at the right place at the right time. James Watson was a newcomer to the scene of Cambridge biochemistry who stole the DNA secret from under the noses of the real experts who had been studying the subject for years. How did he manage that? He had six septiles in his chart, and the harmogram (figure 13.2) shows the huge peak in

---

*Watson's time of birth is recorded on his birth certificate as 1.23 a.m. in Chicago, Illinois (source: UK Astrological Association's data section).

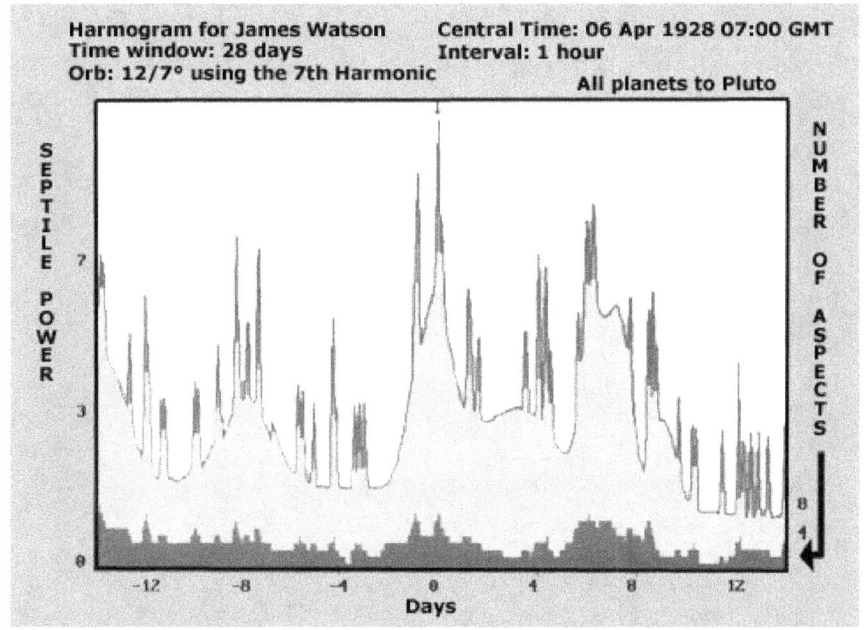

Figure 13.2: 7th harmonic harmogram for birth of James Watson, ±14 days.

septile power over his birth. This helped him to follow — his lucky star.

References

1. Koestler, *The Sleepwalkers*, 1959, p.306.

2. Max Caspar, *Kepler*, Dover 1993, p.106.

3. 1602, letter by Kepler to Fabricius, quoted in Occult & Scientific Mentalities in the Renaissance, Ed. Rosen, CUP 1984, p.255.

4. Fred Hoyle, *Nicholas Copernicus, an essay on his life & work* 1973, p.11.

5. W. J. Tucker, *Your Horoscope and the Fixed Stars*, 1979, p.231.

6. *Aubrey's Brief Lives*, John Aubrey 1813, Penguin 1987, p.18.

7. See the author's Neptune's Discovery webpage at http://www.dioi.org/kn/neptune/index.htm, and for further elucidation, 'A Hiatus in History, The British claim for Neptune's co-prediction', *History of Science*, March & Sept. 2006, vol. 44: http://www.dioi.org/kn/neptunestory.pdf.

8. A. R. Wallace, *My Life*, 1908.

9. Loren Eiseley, *Darwin and the Mysterious Mr X*, 1979, p.27.

10. Dennis Elwell, *Cosmic Loom, the New Science of Astrology*, 1987, 1999, p.150; a noon chart is there given for the 'Origin' publication (p.152).

11. N.K., 'Antares and Aldebaran: Stars of conflict' *Mountain Astrologer*, October 1995, pp.83–85; further discussed N.K.'s 'Interface: Astronomical Essays for Astrologers,' Ascella 1997, Ch.6.

12. Adrian Desmond, *Huxley: Evolution's High Priest* 1997, p.239.

13. Eric Morse, *The Living Stars*, 1988, p.84.

14. Richard Tarnas, '*Uranus and Prometheus*', The Astrological Journal, June 1990, p.151 (c.f., Tarnas, *Prometheus, the Awakener* Woodstock, Spring Pubs., 1995).

CHAPTER 14

Descartes' Dream

'Let the philosopher always be the servant and scholar of Inspiration, and all will be happy.'

William Blake

One should expect some theory of celestial influence to apply to Eureka moments, because through them the universe has been unfolding and blossoming into new forms. They were the seed-beginnings, from which new things have germinated, as did not exist before. Let us hope that, in the future, scientists will take more care to record the precious moments, when new vision dawns. After all, their inspirations have been as much a part of the universe's unfolding as anything else. In their admirable concern for molecules, neurone circuits and distant galaxies, scientists have greatly lost sight of their own psyches as creative agents.

We have here demonstrated that historic eureka moments are not randomly distributed in time. They are attracted by certain harmonic patterns in the heavens as they are likewise repelled by others. We have sought out these *genesis-moments*, when inspiration struck. For their timing, the mould-breaking, prime-number aspects of quintile and septile proved to be the key. We divided the moments into two distinct groups, eureka- and invention-moments, belonging to the somewhat separate streams of the histories of science and technology respectively, and hope to have defined the groups clearly enough for readers to agree, on the whole, concerning their members.

The latter group involved more perspiration than inspiration, and were moments of energy and action, when inventions worked. They contained a huge excess of trine aspects; but, also, the septiles reappeared. These aspects are associated with the numbers 3 and 7 — numbers traditionally associated with deity, as in the seven Days of Creation and the Holy Trinity.

Overtones of these basic harmonics also showed up. The 14th harmonic — i.e. half-septile aspects, multiples of 25.7° — was strongly present in the I-moments, at least as strongly as the seventh; while the fifth harmonic was in deficit, as too were its 'overtones' 10th and 15th harmonics, tending to suggest that as moments of achievement they did not require the creative mental activity in the same way as did the E-moment group. Thirteenth-harmonic aspects were strongly in deficit for the I-moments, which is of interest considering how much is spent by our society in avoiding this lunar number, missing out the 13th floor of buildings, etc. Occasionally, other harmonics seemed to be relevant: *ten* (the decile aspects) was vital for the DNA E-moment, as was *four* for the invention of quaternions, when the 'square' pattern of the heavens dramatically peaked as its fourfold algebra was found, and *two* (strong oppositions) when the first stored-program computer started up, with its diodes a-buzzing and its switches a-clicking on-off.

Astrologically, the charts of great inventions were of visual interest, in their *gestalt* or pattern formed by their aspects. In this respect, they formed a contrast with the E-moment charts, which, depending as they do upon more minor or non-Ptolemaic aspects, were of no interest visually. The barometer, the hot air balloon, anaesthesia in surgery, powered flight, lightning conductor, the jet plane, thermonuclear device, laser beam and computer — these are all memorable charts. As astrologers use the birth moment and not (in general) that of conception, so likewise, more seemed to be expressed by the chart of the I-moment than that of the Eureka insight, when the idea was first conceived.

As regards what the different harmonics *mean*, an interesting remark was made by ISAR President David Cochrane: 'The higher harmonics describe more internal, less conspicuously evident traits in the outer world. Lower harmonics describe more externalised traits.'[1] The eureka research tends to confirm this important thesis, with the

Chapter 14. Descartes' Dream

more concrete I-moments being associated with lower-number harmonics than we found for the more private E-moments. If that generalisation is valid, then it in turn helps to validate the concept of harmonics, as being a useful way of describing celestial aspects.

The same hypothesis was tested on the natal charts of the Eureka-scientists as well as those of the E-moments. This could not be done for the I-moments as our group of inventors of known birth times was even smaller than that of the Eureka scientists, and this despite the fact that the I-moment group was considerably larger that the E-moment group. The group of non-eureka scientists, i.e. those who became famous without any memorable moment of inspiration, can be considerably enlarged as more recent data has become available (appendix A), and this confirms what was suggested back in chapter 5, namely that they have just as significant a deficit of septiles and quintiles, as the E-group has an excess. This suggests a real psychological difference between the two groups.

The transit-linkages of these known charts to those of their inventions were instructive: Edward Teller's Mars in relation to the H-bomb chart; Ben Franklin's Uranus connecting up the 'kite' formation for the lightning conductor chart; Theodore Maiman's Uranus and Mars conjoining the Mars and Uranus of the laser beam chart; the shy and dreamy Joseph Montgolfier having his Uranus-opposition return within a fraction of a degree as all Paris watched his wondrous device, the hot-air balloon, ascending into the sky, and the two suns of the Wright brothers forming a precise grand trine with the sun of their powered flight chart. Such charts fulfil two principles required by traditional astrology: that the event has a major destiny significance and that it can be well-timed.

We looked to some degree at the major planetary life-cycles, where our study focused primarily upon those of Uranus. The most creative period of life between the Uranus trine of 26–28 years and the Saturn return of 29–30 years appeared as the 'breakthrough' period, followed years later by the Uranus opposition of 41–43 years as the time of fulfilment and recognition. Here the main thesis used derives directly from professor Tarnas' thesis in his 'Uranus and Prometheus' essay, where he reviewed the archetype of Prometheus and claimed that it fits the biographies of revolutionary innovators. Here we have narrowed his orbs slightly and focused upon persons who would be called 'scientists'.

Consistent evidence appeared concerning the *shape* and *optimal orbs* of the celestial aspects involved. The applying side of the aspects, coming up to the moment of exactitude, featured more strongly than did the separating side (after exactitude) in the event charts. This was a large effect, of almost thirty percent overall. The result is paradoxical, because an applying aspect is one which has not yet happened. Only after it has chimed, one might suppose, would its influence tend to be exerted. However, astrologers in fact assume, as they always have done, that the applying half of an aspect is more influential than the separating side. Thus the accepted view is confirmed by these studies. As regards the orb for trines, what we called the 'Addey orb' of four degrees ($12° \div 3 = 4°$, in this case) seemed fine, while for the septile it was too large and one degree was about optimal, as denoting the zone of strongest influence.

A Stove at Ulm

Let us go back in time to a key event in the formation of the scientific world-picture, as involved something like a Eureka-moment. Three and a half centuries ago, in war-torn Bavaria, a mercenary soldier named René Descartes experienced a series of remarkable dreams. From them, there developed his view that the world resembled a machine, from which the human spirit was a thing apart. He came to view animals as automata, and humans likewise, except that they had souls and minds in the pineal gland of the brain. 'The modern world, our world of triumphant rationality, began on November 10, 1619, with a revelation and a nightmare ... '[2] This viewpoint dawned on Descartes while he was huddled 'inside' a stove (as he described it) on a cold winter's night at a place called Ulm:

> Descartes conceived his system not, as one might expect, as a mere *calcul*, but with a sense of foreboding and ecstasy. We are bewildered to learn that a mathematical genius should conceive his philosophy with the sense of haunting personal drama. Yet this is precisely what happened. There was one memorable night (November 10, 1619) which Descartes always regarded as the turning-

point of his life. He gives us many indications that his *méthode* was then in its incubation period. On the day preceding that night of dreams he was in a state resembling inebriation. The *scientia mirabilis* seemed within grasping distance. He must have been in that phase close to parturition which we know from the self-observations of great mathematicians, composers and painters. He was aware of the global character of his discovery. Indeed his *scientia mirabilis* seemed a matter of limitless relevance ...

When he woke up from his famous dream, at the birth of his method, possibly shivering at the stove in Ulm from the interstellar coldness of a Cartesian system, he vowed a pilgrimage to Our Lady of Loretto.[3]

Through these dreams he arrived at his 'vision of the unification of all science'. Psychoanalysts have examined Descartes' dreams of that night in great depth, and philosophers have discussed his philosophy to great tedium, but our concern must be with the rather lunar septiles that then chimed.

Figure 14.1 shows a harmogram of the moment. The scale along the base indicates a mere three septiles over the night in question. However, the 'septile power' score (as was discussed in chapter 12) indicates that all three became exact synchronously at midnight, giving a forceful expression of septile power. Two out of these three are lunar, appropriate for a dream-experience — even though, as we have seen, the Moon is normally far from being the most important sphere where E-moment septiles are concerned. Descartes was a character who could sit in front of his fire wondering if he was dreaming, because during his dreams he could appear to himself to be before his fire, in a manner that seemed just as real to him as waking life.

Descartes' published his meditations in June of 1637, as his *Discourse on Method*. The Cartesian precision grid was laid down, the world split apart into subjective and objective, inner and outer, and old subjects like astrology and alchemy were demoted into being 'bad knowledge.' Matters concerning the human psyche became 'merely' subjective. This magnum opus appeared at a time within one degree of Descartes' Uranus-opposition return, as Dr Tarnas has observed.[4]

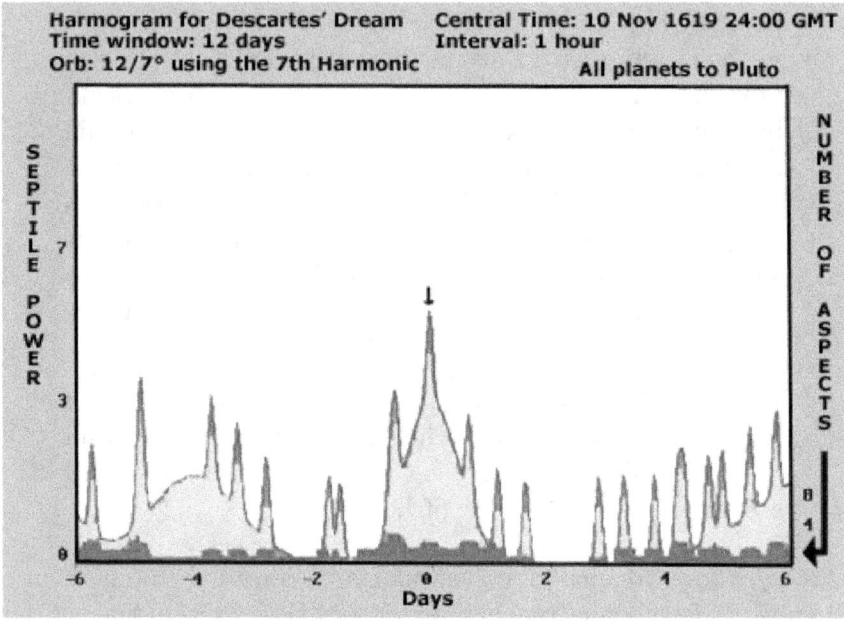

Figure 14.1: Septile power harmogram for Descartes' dream centred on midnight, 10th/11th November 1619 ±6 days.

Between the conception of his philosophy at Ulm to its birth when his 'Discourse' appeared, was a twenty-eight year period of gestation. One is reminded of Charles Darwin who had his moment of illumination during a period of strong septile activity (September 1838), and then published his magnum opus twenty-one years later (November 1859) over a major Uranus transit.

What existed in the world, Descartes averred, was matter and motion, and that was all. That was fine for the newly-burgeoning science of mechanics but it created a problem in accounting for living things and the process of birth. The followers of Descartes (who were called Cartesians) could not credibly account for the genesis of new forms in the birth process. Descartes said that only those things which could be 'clearly and distinctly conceived' should be accepted as true. Physicists may have liked this, but taken in a general sense it cuts one off from the unfathomable. At human birth, an image of the cosmos is, to quote Kepler, 'potentially engraved' upon the soul of the newly-born child. This is a moment of deep mystery. Women believe in astrology more than men, which has often been

Chapter 14. Descartes' Dream

presented as a mental weakness or tendency to superstition on the part of womankind. Rather, we ought to see that it is their experience of the birth process which enables them to apprehend the miracle of that moment, which human reason cannot fathom. A Hermetic science needs to be born, able to reach a little deeper than the finespun patterns of the rational mind.

In the summer of 1619, Kepler's *Harmonice Mundi* rolled off the press, while he was living in Linz. He had been obliged to leave Prague as war was looming. In the following year, Descartes was present at the Battle of Prague, and in November 1620, he marched into Prague with the Catholic Hapsburg victors. Thus one era gave way to another. The difference in world-view between these two, Kepler and Descartes, concerned whether the universe was alive or dead, vitalistic or mechanistic.

We have derived our explanations of how astrology works — to the extent that we ask readers to believe in such matters — from the work of Kepler.[5] He portrayed the deity as tending to realise those things which accord with solar system geometry at any given moment. For example, we saw how the DNA-helix idea appeared at the time of high decile-power, its helix-coil being tenfold. The solar system angles were then in a tenfold mode with regard to the Earth. The DNA moment also had an inspirational tenor for James Watson, shown by its peak in septile power. The septiles defined the E-moment to within an hour, or so. Thus, this moment in time had its own special potency.

Numbers and geometry were causative agents, in Kepler's view, and an example he discussed — and was the first to discuss — was the sixfold structure of the snowflake.[6] The structure of ice crystals determines that pattern, a molecular lattice makes the sixty-degree angle. Likewise, we may see how the fourfold structure of carbon establishes the tetrahedral structure of a diamond, while graphite, in contrast, has a threefold bonding, where all the carbon bonds lie in the same plane. Here we may see the different functions of two numbers, three (trine) and four (square). Graphite is used for movement and lubrication, and is thus opposite in nature to the diamond. Seven was related to the totality of things, for example, a textbook on crystals describes the seven general types of crystal structure, though no crystal can display a sevenfold form.[7]

Such principles were discussed in *The Geometry of Meaning* by the US philosopher Arthur Young, a work which opens with the promising words 'All meaning is an angle.' Young explored the difference between the numbers three and four from a view congenial to that taken here. As the inventor of a helicopter and founder of the Institute for the Study of Consciousness at Berkeley college, he was no mere armchair theorist.[8]

Our survey has sought for a celestial matrix in accord with some fairly Pythagorean concepts. Aspect-patterns appeared as a key to the living force expressing itself through the birth maps of the astrologer. In contrast, we have not been concerned with the time-honoured 'wheels' of astrology - the zodiac of the year, the houses of the day, and progressions of a life's unfolding. This new millennium is time for a synthesis, in which astronomy and astrology come together, appearing as two sides of a coin.

What Matters

Nowadays, physicists discern a twelvefold structure in matter. This remarkable new synthesis arrived with the Great Conjunction of 1993, between Uranus and Neptune, after which high-energy physicists concluded that the entire physical universe was comprised of at most twelve types of matter-particle.

As early as 1979, a popular work on subatomic physics found that:

> In all, there are roughly twelve conservation laws ... These are the laws of symmetry.[9]

Some decades earlier, John Addey had observed that

> ... the number twelve relates to the 'mundane' order of things ...[10]

Twelve is a number that signifies completion, and a stable structure. This theory emerged from the giant particle accelerators of Europe and America[11] as the culmination of decades of endeavour. In April 1994, Fermilab in the US announced that it had finally detected the 'top quark,' implying that all matter was made of six 'quarks' and

Chapter 14. Descartes' Dream

six 'leptons', the former being heavier than the latter. These twelve were composed of three 'families'. The normal matter of the universe was composed just of the first such family (two quarks and two leptons), while the others only appeared under special circumstances. In addition, there seem to be another twelve particles that bind together these matter-particles, which correspond to the forces that enable interactions to take place: photons, gluons and so forth.[12][13] There are now two sets of particles each in groups of twelve, with a mirror-symmetry between them.

As there are twenty-four vertebrae in the spinal column, twenty-four hours in the day, and twenty-four letters in the Greek alphabet, so the third millennium kicks off with that number of 'fundamental' particles. Also, as the ancients believed that the world was composed of four elements, so physicists believe that there exist just four different types of force in Nature (each having their own force-particles). Do four matter-particles now compose the physical world? The stable structure of matter expresses itself through the numbers four, twelve and twenty-four. It is becoming ever clearer that particle physicists need a grounding in Pythagorean number-theory.

In the seventeenth century, the idea of celestial influence had become radically de-legitimised partly because of the new atomic theory. The word 'atom' means 'uncuttable', from the Greek *a-tomos*. Atomism, as it existed since the time of ancient Greece, has enjoyed a long and fruitful career, until it came to an end in the year 1967. This Greek concept became established as commonsense in the eighteenth century, then became problematic around 1930 with quantum concepts, and finally terminated when, in 1967, single particles of photons and electrons were made to pass through a double-slit and interfere with themselves.[14] They created their own interference-pattern, by passing through both slits at once! Comparable results were later obtained for electrons and neutrons. The notion that such particles were 'uncuttable' was no longer valid. The 'fundamental' particles were no longer solid: matter could not be made of anything resembling billiard balls.

If we are claiming that the *macrokosmos*, the planets, affects human consciousness, as we are, then such shifts in perspective have to be profoundly significant. The notion is no longer in such a direct violation of the concepts of science, as has appeared to be so in recent centuries. We seem to be moving towards a new synthesis which may not yet have happened. Science seeks for the pattern in matter: pattern from *pater*, the father, and matter from *mater*, mother. The definitions of things change, and what seems evidently possible alters in our experience. Exiting from atomism, we have arrived at a quantum-entangled universe.

The concept of 'entanglement' arose, to describe how pairs of particles as can be miles apart somehow know what each other is up to. Quoting from a *New Scientist* article of 1998:

> Non-locality cuts into the idea of the separateness of things, and threatens to ruin the very notion of isolation. To isolate an object we ordinarily move it a long way away from anything else, or build impenetrable walls round it. But the link of entanglement knows no boundaries. It isn't a cord running through space, but lives somehow outside space. It goes through walls, and pays no attention to distance.

Quantum entanglement 'maintains links between distant regions, and keeps the whole universe coherently connected'.[15] This *is* action-at-a-distance. The beautiful notion of 'unbroken wholeness' proposed by the quantum physicist David Bohm (at Birkbeck College, London, in 1974) seems to be growing in importance.

Einstein famously disliked what he called 'spooky action-at-a-distance' in quantum mechanics, and objected to 'using telepathic means as the current quantum theory alleges'.[16] His scruples weren't resolved until 1981, when an experiment definitely established that such 'spooky action-at-a-distance' did normally and commonly occur.[17] Thus we have exited from the Cartesian universe in which forces were only transmitted by things touching each other.

Now and then the scientific community reacts against astrology. In 1994, tones of shrill hysteria were heard from the journal 'Nature's' editorial:

> Each and every horoscope is, by denying the objective view of the planets, an attack on the probity of science.[18]

This was the case, it explained, not merely 'because there is no evidence that the position of the planets can affect human behaviour',

Chapter 14. Descartes' Dream

but because 'those who peddle horoscopes do so on an explicit set of statements about the real world that *cannot* be correct' (my italics). The editorial intoned against the lurking threat of 'anti-science,' in which religion and astrology found themselves placed as bedfellows; a view hitherto only expressed in America.

There remains an abyss between the two traditions, with astrology based upon a concordance and interlinking between earth and sky that is inherently meaningful, while the procedures of science are material, analytical and deductive. The former finds proof in the test of experience, that the client comes back, and has been coming back for two and a half millennia, while the latter finds proof from the readings of apparatus. The *astro-logos* of a horoscope has an inherently circular meaning that involves symbolism, whereas the logic of science is linear, it will come to a point, and has no concern with symbols. The contrast is very much one between left- and right-brain modes of thought, as well as being nowadays very strongly gender-polarised.

Looking back into the past, we see a rift or fault-line as developing geographically, perhaps around the line of East 30° in longitude, just West of the Nile: astrology developed to the East of this, from its birth in Chaldea (now Iraq) and then journeyed Westwards into Syria and Egypt. At Alexandria the great Ptolemaic synthesis took place, absorbing components of Greek thought and religion, though Greece had *no formative role* in its development. Today's science developed due West of this line, passing from Greece into the Arab lands of North Africa and thence into Europe during the Renaissance.

The Greek notion of *Kosmos* involved a proportion between man and the solar system, and it is that sense of proportion which we have here explored. The old Greek word *Kosmos* also meant beauty (as in *cosmetic*), and can we still experience beauty in finding a proportionality between man/woman and the revolving solar system? It seems that there are certain windows in time, due to relationships being formed between the bodies of the cosmos, when the imaginative and intuitive faculties can be awakened. Up till now this has happened unconsciously. The new millennium sees the possibility of a more conscious participation, so that Time in our experience comes to have a different significance.

It is beneficial for scientists to ponder the harmony implied by the concept of aspects. As an electron orbit is defined by the limited whole-number ratios of quantum theory, so likewise only certain integral divisions of the circle generate effective aspects. The specialist who has lost the ability to think about the whole of things is a dangerous person. Perhaps we need to hear anew the voice of the astrologer, because she never accepted the axiom that the universe is a non-living thing there to be exploited by man, nor has she forgotten, as the universities of learning have so shamefully, that the human spirit has an innate connection with the stars above.

Poets appreciate this. Sir William Hamilton, the Irish mathematician and astronomer, asked Wordsworth for his views about science.[19] The poet had published some scathing verses upon the subject ('Philosopher, O fingering slave/ He that would peep and botanise upon his Mother's grave ...We murder to dissect'). Did this mean that Wordsworth was quite opposed to science? He visited Ireland to meet the astronomer in 1829, and in the grounds of the Dunsink Observatory they discussed how the intellect and the imagination should be related.

Wordsworth told Hamilton that he would not be opposed to a science that 'raised the mind to the contemplation of God in works', though he did object to a science applied only to material uses of life, because he saw this as one which 'waged war with and wished to extinguish imagination.' In Book IV of the *Excursion*, Wordsworth envisioned a science freed from the 'false conclusions of the reasoning power.' He there presented an allegory of a child listening to the sound of the sea in a shell, in which the shell was 'the universe itself', resounding to the 'authentic tidings of invisible things, of ebb and flow and ever-during power.'

The scientist who sought to reveal this harmony would indeed be worthy of his calling:

> ...*Science then*
> *Shall be a precious visitant; and then,*
> *And only then, be worthy of her name:*
> *For then her heart shall kindle; her dull eye,*
> *Dull and inanimate, no-more shall hang*
> *Chained to its object in brute slavery;*
> *But taught with patient instinct to watch*
> *The processes of things, and serve the cause*

Of order and distinctness ...

References

1. D. Cochrane, *Astrology in the 21st Century* Florida 2002, p.94.

2. *Descartes' Dream, The World According to Mathematics*, P. Davis and R. Hersh, N.Y., p.1

3. K. Stern, *The Flight From Woman*, N.Y. 1965, p.105.

4. R. Tarnas, 'Uranus and Prometheus', *The Astrological Journal*, Sept/Oct 1989, p.245.

5. Quotes from 'Kepler's Astrology, excerpts' by Ken Negus, 1987 Princeton, N.J. There is a fine essay on 'Kepler: astrology and mysticism' by science historian Edward Rosen in *Occult and Scientific Mentalities in the Renaissance* Ed. Vickers CUP, 1984.

6. C. Schneer, 'Kepler's New Year Gift of a Snowflake', *Isis* 1960, 51, pp.531–545.

7. A. Clark, *Minerals*, Hamlyn, 1979, p.8: 'the seven crystal classes'.

8. A. M. Young, *The Geometry of Meaning* Robert Briggs Associates, CA, 1976.

9. Gary Zukav, *The Dancing Wu-Li Masters*, 1979, p.260.

10. Addey, *Harmonics in Astrology* 1976, p.101.

11. *Back to Creation*, CERN Publications, Geneva 1991, p.5.

12. Gordon Kane, *The Particle Garden, our universe as understood by particle physicists*, 1995, p.116. The twelve matter-particles are called 'fermions' and the 'glue' particles, 'bosons'.

13. Joseph Schwartz, *'The Creative Moment, How Science made itself alien to modern culture'*, 1992, p.166.

14. Fred Alan Wolfe, *The Dreaming Universe*, 1994, p.181.

15. Mark Buchanan, Why god Plays dice, *New Scientist*, 22 August, 1998, pp.27–30; see also D. Bouwmeestr et. al., *Nature*, 11 Dec 1997, pp.575–579, 'Experimental Quantum Teleportation'.

16. A. Pais, *Subtle is the Lord*, Oxford 1982, p.440.

17. Photon-pair experiment by Alain Aspect & colleagues at Paris, 1981.

18. Editorial by John Maddox, *Nature*, 368, March 1994, p.185. Let's here note that Nature published its famous 'Carlson experiment', ('A double-blind test of astrology') in 1985, the most widely — and consistently-cited — negative-result test of astrology. However, a re-evaluation has showed that, contrary to the initial

judgements and conclusions, the astrologers did in fact then obtain weak positive results — under quite adverse circumstances (Dr Suitbert Ertel, 2009, 'Appraisal of Shawn Carlson's Renowned Astrology Tests', Journal of Scientific Exploration, 23, pp.125–137). For discussion, see Robert Currey, 'U-turn in Carlson's astrology test?' Correlation, 2011, 27, pp.7–33. So, to claim that scientific tests show that astrologers cannot recognize the charts of people put in front of them is no longer true. One must surely concur with Robert Currey, that Nature only published the study because it believed it had yielded a negative result.

19. T. L. Hankins, *Sir William Rowan Hamilton*, 1982, p.103.

Appendices

APPENDIX A

Birth data

Eureka scientists (n=16)

GMT TIMES ARE GIVEN IN BRACKETS, sources are cited last
ABC = American Book of Charts;
Gauquelin = 'Birth and Planetary Data Gathered Since 1949', Vol. 2 Men of Science 1971 by M & F Gauquelin;
AA = the Astrological Association of Great Britain's data collection.
Birth dates of scientists not mentioned below will usually be found in Asimov's BEST.

Brahe, Tycho: 10.47 a.m. (09:55hrs) 24/12/1546 NS, (14/12/1546 OS) Lund, Sweden, 55N40 13E30, ABC.

Galileo, Galilei: 3 p.m. (14:18hrs) 26/02/1564 NS (15/02/1564 OS), Pisa, Italy, 43N43 10E24.

Kepler, Johannes: 2.30 p.m. (13.54hrs) 27/12/1571 OS, (06/01/1572 NS), 47N36 7E39.

Davy, Humphrey: 5 a.m. (05:22hrs) 17/12/1778, Penzance, UK, 50N07 05W33.

Pasteur, Louis: 2 a.m. (01:38hrs) 27/12/1822 NS, Dole, France, 47N08 5E30, Gauq.

Roentgen, Wilhelm: 16 hrs LMT (15:31hrs) 27/03/1845 NS, Lennep Rhineland, Bordini.

Edison, Thomas: 11.30 p.m. (04:30:25hrs 12 Feb), LMT 11/02/1847, 41N18 82W36.

315

Becquerel, Henri: 3 p.m. (14.51hrs) 15/12/1852 NS,
Paris, France, 48N52 2E20, Gauq.

Tesla, Nicola: 'Just after midnight' (00:15hrs) 21/07/1856 OS
(10/07/1856 NS, 44N35 15E19; ABC.

Einstein, Albert: 11.30. a.m. (10:50hrs) 14/03/1879 NS,
48N24 10E0; Ebertin.

Heisenberg, Werner: 4.45 a.m. MET 5/12/1901 NS,
Wurzburg, Rodden.

Fleming, Alexander: 2 a.m. (02:00hrs GMT) 6/08/1881 NS,
55N37 4W18, ABC.

De Broglie, Louis: 1 a.m. (00:50:40) 15/08/1892 NS,
Dieppe, 49N56 1E05; ABC.

Fermi, Enrico: 7 p.m. (18:00hrs) 29/09/1901,
Rome, Italy, 41N53 12E30. Gauq.

Townes, Charles: 5 a.m. EST (10:00hrs) 28/07/1915,
Greenville, South Carolina, 34N51, 82W24 Gauq.

Watson, James: 1.23 a.m. CST (07:23hrs) 6/04/1928,
Chicago, Illinois, 41N53 87W38, A.A.

Further:-

- **Edison's** birth time was taken as 11.30 p.m. local time on 11 Feb 1847 in Milan, Ohio (5:00hrs GMT, 12 Feb 1847).
  This time is given by
  a) De Lascaut, who is normally regarded as reliable;
  b) the Circle Book of Charts;
  c) Cirbels;
  d) T. Ring, *Astrologische Menschenhunde* Vol. 3, p.539;
  e) the UK journal, 'The Astrologer' in its Feb. 1890 issue,
  which gave Edison's birth data as 11.33 p.m. local time on the above date. Its authority cited was a 'professor Chaney' who had obtained it from Edison's father. The claim that Edison was born during a blizzard at 3 o'clock in the morning (in 'Thomas A. Edison' by M. C. Nerney) sounded like the sort of myth likely to develop around the birth of an electrical genius (Fowler's Compendium gave 1.30 p.m., a printing error for 11.30 p.m.).

- **Galileo's** birth date is often taken as 15th Feb, 1564 OS.
  However, it should be a day later, 16th Feb, 3 p.m. GMT 1564 OS;
  (N.K. *Interface, Astronomical Essays for Astrologers*, 1997, 'Galileo's Birth-date') as shown by two horoscopes recognised as drawn up by

him for his birth at the National Library, Florence. (The two differ by merely half an hour.)

- **Tesla** was born just after midnight on the 9th July 1856 in Croatia. Much of the Balkans was still using the Julian Calendar, but sources (e.g., Campion's *Book of World Horoscopes*) generally concur that for Tesla's birth date, the modern Gregorian calendar was used, and his birthday celebrations in New York in his later years confirm this.

- Eureka scientist **Ben Franklin** is not included, because his birth data of 4h 44m LMT, 17 Jan 1706 NS Boston, is cited by Rodden as 'dirty data' i.e. having no known source;
  likewise **Niels Bohr**, who has a slight claim to being a eureka-type, 0h 50m LMT, 7 Oct 1885 Copenhagen, has the same Rodden rating.

****

The Non-Eureka group (n=20)

Copernicus, Nicolaus: 4.48 p.m. (15.34) 19/02/1473 OS
    53N1, 18E37, Gauquelin.

Vesalius, Andreas: 5.45 a.m. LMT (05.27) 10/01/1515 NS.

Hooke, Robert: 12 noon 28/07/1635,
    Freshwater, IoW; AA.

Flamsteed, John: 7.16 a.m. (07.22) 19/08/1646
    Derby 52N55, 1W29; AA.

Halley, Edmond: 10.30 a.m. 29/10/1656 OS
    London.

Lavoisier, Anton: 9.30 a.m. (9.20 GMT) 26/08/1743,
    Paris, AA.

Bode, Johann: 1.15 p.m. (12.35 GMT) 19/01/1747,
    Hamburg, AA.

Brewster, David: 12.15 p.m. LMT 11/12/1781
    Jedburgh, Scot.)

Leverrier, Urbain: 10 a.m. 11/03/1811,
    49N07, 0W04.

Huxley, Thomas H: 9.30 a.m. 4/05/1825
    Ealing 51N30, 0W15. AA.

Crookes, William: 5.30 p.m. LMT 17/06/1832,
London.

Curie, Marie: 10.36 a.m. GMT, 7/11/1867
Warsaw.

Hahn, Otto: 11.45 a.m. LMT 8/03/1879
Frankfurt.

Joliot-Curie, F: 9 a.m. 19/03/1900
Paris.

Pauling, Linus: 7.26 a.m. PST 28/02/1901
Portland, OR.

Segré, Emilo: 12.00 p.m. (11 a.m. GMT) 1/02/1905
Rome.

Bethe, Hans: 20.45hrs 2/07/1906
Strasburg.

Seaborg, Glen: 4 a.m. CST (10 a.m. GMT) 19/4/1912
Ishpeming MI.

Ehrlich, Paul: 15.20hrs CST 4/05/1932
Minneapolis.

Sagan, Carl: 22.05hrs 9/11/1934,
New York AA.

Asimov scientists of reliable birth date with English-language biographies, found recently, i.e. not in chapter 5, are, in addition to the above: August Kekulé, Baron Georges Cuvier, Justus von Liebig, Pierre Laplace and Alfred R. Wallace.

Appendix A. Birth data

The Inventors with known birth times

Baird, John L: 8 a.m. 13 Aug 1888 Dunbartonshire, 56N1 4W44.

Maiman, Theodore: 8 a.m. (16:00hrs GMT) 11 July 1927, Los Angeles, (from Maiman's secretary).

Marconi, Guiglio: 9.15 a.m. (08.25) 25 April 1874, Bologna. Gauq.

Wright, Otto: 22.36hrs 19 Aug. 1871 Ohio, US.

Bell, Alexander: 7.12 a.m. 3 Mar 1847 Edinburgh.

Morse, Samuel: 10.30 a.m. (1 p.m.) 27 April 1791 Mass. US.

Glaser, Donald: 11.30 p.m. 21 Sept 1926 Cleveland Ohio.

Other scientists given in text

Hamilton, William: 12 midnight, 3/4 August, 1805, Dublin.

Leibniz, Gottfried: 6.12 p.m. (17.22) 1 July 1646 NS, Leipzig.

Lowell, Perceval: 7.45 a.m. LMT 13 Mar 1855 42N21 71W4 (Mark Edmund Jones).

Turing Alan: 2.15 p.m. GMT 23 June 1912.

Wallace, Alfred: 2.30 a.m. (02.41) 8 Jan 1823, Usk, Monmouth 15N43, 2W54.

Asimov E-scientists with lost birth times

C. Darwin, M. Faraday, A. Fleming, D. Gabor, A. Galle, L. Galvani, J. Gutenberg, E. Hubble, L. Meitner, D. Mendeleef, H. Ørsted, W. Perkin, M. Planck, W. Ramsay, I. Semmelweis, L. Szilard, C. Tombaugh, J. Watt, A. Werner.

Other Asimov scientists with known birth times

Jerome Balard, Paul Bert, Edouard Beneden, Louis Cailletet, Owen Chamberlain, Louis Chardonnet, Armand Fizeau, Francois Grignard, Otto Hahn, Pierre Janssen, Jacobus Kapteyn, Hans Krebs, Max Laue, Joseph Lebel, Henri Le Chatelier, Joshua Ledeberg, Abbe Georges Lemaitre, Giulio Natta, Jean Baptists Perrin, Auguste Piccard, Francois Rauolt, Paul Sabatiar, Emilio Segré, Jean Stas, Herman Staudiger, Georges Urbain, Heinrich Wieland, Robert Williams, Karl Ziegler.

APPENDIX B

Statistical procedure

THE CHI-SQUARE TEST REQUIRES THAT the units counted be independent of each other. That is to say, the appearance of one septile or quintile should not affect the likelihood of another taking place. That is largely the case, but not entirely: if two planets are in a quintile relationship to another, then it is likely that they will themselves form another quintile or biquintile aspect between themselves. The chi-square may tend to give an exaggerated significance level, since the units measured (aspects) are not wholly independent; while the t-test requires an approximately 'normal' distributions of the data, whereas a distribution of the number of quintile and septile aspects per chart, or in a group of fourteen charts, is skewed and discontinuous. Instead, an empirical method had to be used.

Monte Carlo method of probability assessment

To estimate the significance of the quintiles and septiles present in the fourteen natal Eureka charts, a control group was generated. The method advocated by Prof. Arno Mueller was used. For aspect frequency studies, he explained, a control group should be generated by taking the same proportion of births per decade as the original sample. This approach was necessary because 'There can not exist expectancy values of general validity to be applied to all samples in a simple manner.'* In our sample, the charts are unevenly distributed over four centuries of time, with none born in the seventeenth century and seven from the nineteenth century.

---

*Prof. Arno Muller, 'Comments on Astronomical and Statistical Problems with Astrological Aspects', *Astro-Psychological Problems* (Paris, Ed. Francoise Gauquelin), May 1986, 14–16. For further discussion of estimating the expected aspect frequencies, see N.K. 1995.

The computer created a group of fourteen charts, randomly selecting one from each decade centred on one of the natal E-group. Then, it computed the Q+S score from that group. This procedure was performed one hundred thousand times, taking the period 1800–2000 and selecting noon times at ten-day intervals. It then counted the groups where the Q+S score exceeded a certain value. Our initial score for this group was 88 Q+S aspects, and the computer found 329 of these. This gave an empirical probability of just over 1 in 300, which is almost indistinguishable from that obtained by the chi-square formula. From this it was concluded that the chi-square probability estimate was quite adequate.

Expected frequencies

a) Empirically generated

For estimating the expected frequency of a given aspect-harmonic, it is preferable to include the conjunction position as Addey recommended. For the third harmonic for example one would then sample 0°, 120° and 270° positions between two planets, plus or minus the chosen orb. If the zero position is omitted, this creates long-period irregularities, as a result of slow movements of the outer planets, and this asymmetry makes a stable expected frequency harder to obtain. We chose to do this because, in sampling a sum of two aspects, one couldn't score the zero position twice.

Noon times were taken for this procedure. As before, Addey orbs of $(12 \div n)°$ were used where n is the harmonic number. Most of the births of the 14 scientists occurred within a period of 110 years, from 1820 to 1930, which was our first sampling group. Some mean frequencies which the computer generated were:

Sampling Period	Mean Frequencies	
Every ten days	Quintile	Septile
1820 — 1930	2.20	2.39
1800 — 2000	2.16	2.41
1500 — 2000	2.22	2.42

The overall means are 2.19 for the quintile and 2.41 for the septile (compare 2.25 and 2.35, respectively, from table 5.2, chapter 5: The Eureka Effect). Likewise, for each of the 36 I-moments, the computer sampled a decade centred on it, taking daily noon values over that decade. Thus, the 1940s were sampled several times, because there were quite a lot of I-moments in that decade. This is the control method recommended by Prof. Muller. It was basically designed for the inner planets, where expected frequencies are far more variable.

b) Theoretically computed

For major aspects to Uranus (chapter 8: Uranus the Awakener) at 5° orb, the expected frequency per chart is found as follows: there are nine planets (including Sun & Moon) to which aspects can apply and a conjunction can fall over 10° of the ecliptic (i.e. ±5°). Likewise for the opposition, but 20° for the trine as there are two possible trine positions, so the overall expected frequency E per chart is given by

$$\frac{9(10+10+20)}{360} = 1.0.$$

Likewise, taking merely two degrees of orb for 36 charts (see figure 10.1, page 222),

$$E = \frac{9x16}{360} \times 36 = 14.4,$$

and for all six Ptolemaic aspects to Uranus at 5° orb,

$$E = \frac{9(10+10+20+20+20)}{360} = 2.0 \text{ per chart.}$$

These closely agree with empirically-computed values, since there is little Sun-clustering effect to distort them — as is however the case for septiles.

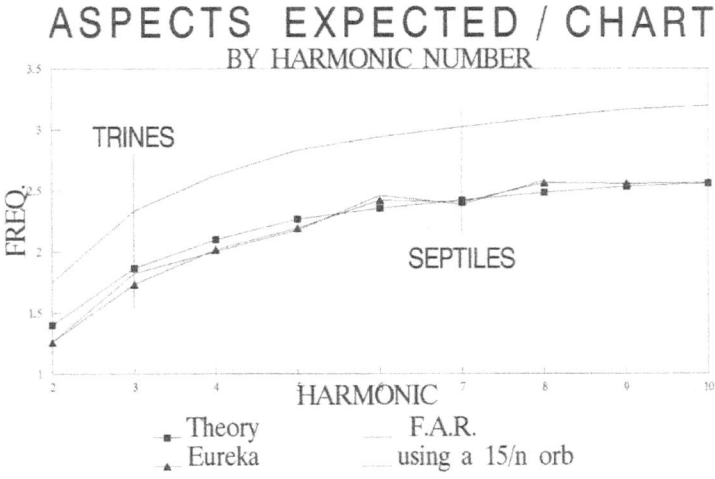

Figure B.1: Aspects Expected chart.

Septile expected values are found as follows: there are a total of 45 planet-pairs between the ten planets (10! = 45), of which three cannot form most aspects (Venus/Sun, Mercury/Sun, Venus/Mercury), therefore leaving 42. So, to a first approximation, the expected value for any harmonic is

$$\frac{42 \times 24}{360} \times \frac{(n-1)}{n}, \text{ for an orb of } \frac{12}{n},$$

ignoring possible aspects between Venus, Mercury and the Sun.

Suppose we want the expected frequency of septiles, plus bi- and tri-septiles. To start with,

$$\text{E (monoseptile per planet-pair)} = \frac{12}{7} \times \frac{2}{360} = \frac{1}{105}.$$

Venus has a maximum elongation from the Sun of 48° and Mercury 28°, so Venus and Mercury can form a mono-septile between them, while there are 42 planet-pairs that can form all septiles, so:

$$\text{P (all septiles per chart)} = \frac{(42 \times 6 + 2)}{105} = 2.42$$

This gives the expected frequency for septile aspects per chart at the given orb.

The graph (figure B.1) compares empirically and theoretically computed aspect frequencies, using the Addey orbs. It compares the expected frequencies per harmonic, as used in Project Eureka, with the expected frequencies per harmonic using *Pottenger's Frequencies in Aspect Research* program, as found by Kevin Hawley. Empirical values are slightly below the expected for the first five harmonics (due to solar clustering) and this difference fades away for the higher harmonics. An additional line depicts the chance-expected values for an Addey orb of $(15 \div n)°$, as some may prefer.

c) Chi-Square

The natal E-group comprised 16 charts with a total of 99 Q+S aspects, where the expected is 73.6, derived from 16 x (2.19 + 2.41), bringing the X^2 value to 8.8 for a net excess of 35%. Tables convert this into the probability-value of 1 in 300. Our earlier-published analysis had only 14 members. The 23 members of the E-moment group have 145 Q+S aspects and 105.8 expected giving X^2, equivalent to

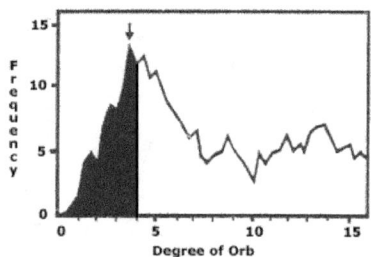

Figure B.2: Trine aspects at Invention-moments.

a probability-value around 1 in 2000 (For E-moment group, see table 5.1: The Eureka Moments (n=23), page 64).

There is an alternative approach to using the chi-square in this context, as involves varying the orb of the aspect. This method estimates

Appendix B. Statistical procedure

the optimal orb. If one scores, say, the frequency of trine aspects as one gradually increases the orb, say for for the 36 I-moments, and compares it with the expected frequency, the varying X^2 values thereby derived will give the graph as shown (figure B.2). Its peak points to the best orb, i.e. that giving the most significant result. As can be seen, this was very close to the 'Addey orb' $(12 \div n)°$ of four degrees.

d) Including the conjunction

It is simpler to compute the theoretically expected frequency of aspects per chart if the conjunction is included. Using an Addey orb, the expected frequency then becomes *independent of the harmonic number*. Also, the aspects between the Sun and Mercury and the Sun and Venus would then no longer be excluded, as they will form the 0° aspect, making the total number of possible aspects 45 rather than 43. For a given harmonic this can be computed as:

$2 \times \frac{12}{n} \times \frac{n}{360} \times 45 = 3.0$, using the 'Addey orb' set at $(12 \div n)°$.

The harmonic number N cancels out from the formula (N.K. 1995, p.293).

Including the conjunction gives more stable frequencies, so that the empirically-determined scores become fairly close to the theoretically computed value. Taking noon times every 5 days over 5 centuries, 1500–2000, the computer obtained the means of 2.96 for the 7th harmonic and 2.93 for the 5th, as mean expected frequencies per chart.

APPENDIX C

The original E-moment list as presented in October 1987

1672	Tycho Brahe sees new star (nova)	3,2
1618	Kepler finds 3rd law of planetary motion	1,4
1774	Priestley prepared oxygen	0,2
1781	Herschel observes Uranus	2,2
1801	Davy makes new metal (potassium)	1,3
1831	Faraday - electromagnetic induction	2,4
1846	Galle sees Neptune	3,5
1869	Mendeleev discovers Periodic table	0,6
1895	Roentgen notices X-rays	5,9
1877	Edison invents phonograph	7,3
1896	Becquerel discovers radioactivity	1,2
1921	Loewi discovers nerve transmission	4,3
1928	Fleming discovers penicillin virus	4,7
1930	Tombaugh sees Pluto	1,2
1933	Szilard realises chain reaction theory	2,4
1942	Fermi switches on nuclear pile	3,3
1947	Shockley finds semiconductor principle	5,2

The published E-moment lists of 1988 and 1996 by M.O. & N. K. had 21 members, as in the chapter 5: The Eureka Effect list but lacking the Hubble & Edison moments, and it also used an earlier (evening) time for Roentgen's X-rays. It scored 54 quintiles and 76 septiles.

APPENDIX D

Lost moments (no dates could be found)

Lost Eureka-Moments (n=10)

GUTENBERG, Johannes, printing press, p.35.

WATT, James was strolling across Glasgow's golf-course, 'on a fine Sabbath afternoon' in May 1765 (can one locate the date?) when the idea for how to work his new steam-engine design dawned upon him.

GALVANI, Luigi in Bologna watched a dissected frog's leg twitch when an electric current was applied: his notebook on November 6th, 1780 described something like this, but he may have made the observation earlier. (Koestler, *AOC*, p.667).

AMPERE André-Marie, electric current induction, p.37.

SEMMELWEIS, Ignac in Vienna had a dream of his germ theory of disease in 1847: 'In the excited condition in which I then was, it rushed into my mind with irresistible clearness ...' He died in an insane asylum, depressed because his views about simple hygiene were rejected.

PASTEUR, Louis in Paris discovered crystal optic rotation in 1848, on 'one of the first days in May.'

WALLACE, Alfred Russell, envisaged the process of biological evolution on a tropical island in February of 1858 — although doubt remains over his account (p.245)

TESLA, Nicola. As the Sun set in Budapest Park one February day in 1882, the idea came to Tesla, how an alternating current dynamo could be made, something everyone 'knew' was impossible. There had to be a way, he felt sure, and for four years he had been grappling with the question. How could a motor make alternating current? He happened to be reciting some lines from Goethe's Faust about the Sun — having his friend Szigeti by his side — when 'the idea came like a flash of lightning

'... The images I saw were wonderfully clear and sharp and had the solidity of metal and stone, so that for many years afterwards my life was little short of continuous rapture.' With a stick he drew his diagram on the ground: six years later he explained to the American Society of Civil Engineers, the new method of electrical power transmission. Soon it was used by all the world. (*Nicola Tesla My Inventions: The Autobiography of Nicola Tesla* 1919, 1981, p.61. Also, *The Electrical Experimenter*, 1919. For the deep originality of what Tesla achieved, see *Wizard, The Life and Times of Nicola Tesla* 1998, pp.24–25).

WERNER, Alfred in Zurich, 1891, awoke from sleep at 2 a.m. having envisaged his co-ordination theory of chemical bonding for organic compounds.

FLEMING, John Ambrose 'founder of electronics' invented the thermionic valve: 'Thinking over the subject intensely, I had in October 1904, a sudden very happy thought.' His assistant performed an experiment, which 'was at once a great success', then a month later Fleming patented his first 'valve'.

GABOR, Dennis, discovered the hologram principle in 1947: 'After I had pondered the problem for a long time, a solution suddenly dawned upon me, one fine day at Easter ...', while sitting down at his tennis club at Rugby. But, no-one could make it work until years later, for that needed laser beams.

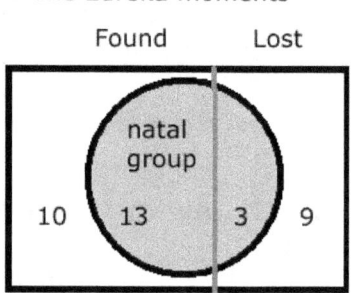

Figure D.1: Lost and found Eureka moments.

The diagram shows the different Eureka subsets. Of the historic eureka-moments, 23 are 'found', i.e. their dates are known, and 10 are 'lost'. N.B., the 'lost' E-moments are believed to be genuine, and are not the same as illusory E-moments, such as those which are attributed in stories to Isaac Newton, August Kekulé and Alfred Russell Wallace (See pp.8,245–6.)

Lost Invention-moments (n=27)

1608: The telescope (Lippershey)
1656: Pendulum clock (Huygens)
1712: Steam engine (Newcomen)
1767: Spinning jenny (Hargreaves)
1800: Voltaic cell (Volta)
1802: Spectroscope (Wollaston)
1832: Dynamo (Pixii, Paris)
1835: Photographic negative (W. Fox Talbot, in August)

Appendix D. Lost moments (no dates could be found)

1837: Electric telegraph (Morse)
1839: 'Daguerrotype' photograph (Daguerre)
1839: 'Velocipede' i.e. the original bicycle (Macmillan, Fyfe)
1852: Gyroscope (Foucault)
1857: Aniline dye (Perkin)
1860: Dynamite (Nobel)
1862: Motor car (Etienne Lenoir)
1865: Antiseptic (Lister)
1880: Hydro-electric power (Lord Armstrong, Cragside)
1883: A. C. Electric motor (Tesla, in May)
1895: Cloud chamber (Wilson)
1904: Thermionic valve (Fleming, in October)
1904: Gyro-compass (Anschutz)
1904: Stainless steel (Brierly)
1909: Neon light (Claude)
1930: Cyclotron (Lawrence)
1932: Electron microscope (Knoll and Ruska, Berlin)
1937: Radio telescope (Reber)
1952: Bubble chamber (Glaser).

Some of these might be findable, e.g.,

1800, **Voltaic Cell**: On 20th March 1800, Volta sent a letter to the Royal Society, London, announcing his voltaic pile, the first electric battery. It would have been constructed shortly before.[1]

1837, **Telegraph**: The electric telegraph was invented independently in Britain and the US. By 1837, Samuel Morse had assembled the main elements of his telegraph design. In October, a request for a patent was applied for, specifying what he intended to patent when it should be in completed form. He brought his plans to an eminent American scientist Joseph Henry, who advised him on design. In the same year a patent for an electric telegraph was taken out in Britain by Cooke and Wheatstone. F. W. Cooke was studying at Heidelberg, where he saw demonstrated some telegraphic experiments with a galvanometer, and realised that he could make a commercial success of it. Professor Charles Wheatstone at King's college had also been experimenting with telegraphic apparatus, so they decided to go into partnership. The telegraph which they patented in 1837 used several wires concurrently. By 1838, they managed with only two wires, and this design was installed on the Great Western Railway in that year, between London and West Drayton.

1857, **Aniline Dye**: W. H. Perkin stumbled across the method of making aniline dyes in his Easter vacation 1856, then by June 1857 he had set up the Greenford Green factory with his father's assistance, and 'Tyrian purple' was being used in London dye houses within 6 months. It soon became highly fashionable. In London of 1906, the 'Jubilee of the discovery of mauve' was celebrated.

1865, **Antiseptic 1865**: In the wards of the Glasgow Royal Infirmary, as in every other hospital in the mid-nineteenth century, nearly half of the subjects of major surgical operations died from the so-called 'hospital disease.' Surgeons changed into bloodstained aprons before operating. Doctors vaguely believed that there was something in the air which caused operation wounds to putrefy. Lister was advised by a colleague to read Pasteur's new observations, and wondered if the infection could be a result of invisible micro-organisms.[2]

Lister used carbolic acid on a compound fracture in March 1865, but without success. On the 12th August of that year, a boy called James Greenlees, aged 11 years, was admitted to the Glasgow Royal Infirmary with a leg fracture. This was 'the first of a series of eleven cases of compound fracture treated between that date and April, 1867'. That was the first hospital use of an antiseptic. The wounds healed rapidly and a new era of cleanliness dawned in hospitals.

1910, **Neon Light**: It was developed by the French physicist Georges Claude, and displayed for the first time at the Paris Motor show on 3rd December 1910 (SBF). It shone red when an electric charge was put through it. Nicola Tesla is alleged to have invented a fluorescent light in 1900, but Tesla I-moments are all lost.

1914, **Stainless Steel**: Harry Brearley was a plain-speaking Yorkshireman who invented stainless steel, around 1912. He found it while studying problems related to ordnance. He sent samples of the new steel — high in chromium, and cooled by a certain process — to various firms, to no avail, then eventually he tried a Mr Ernest Stuart, the cutlery manager of Messrs R. F. Mosley, in June or July of 1914. Stuart made some knives that worked, from the stainless steel. 'So far as the initial use of stainless steel is concerned, Mosley's are the firm to whom credit is due,' wrote Brearley. Fierce legal battles broke out once it became evident that this new steel really worked, with attempts to divest Brearley of all credit. An American firm came to his rescue. Brearley told the story in the Sheffield Daily (23rd February, 1924). The loss of all relevant dates, including that of Brearley's birth, is regrettable.

1937, **Radio Telescope**: A student at Illinois Institute of Technology called Grote Reber built the world's first radio telescope in his back yard, in September of 1937. He was following the pioneering work of Jansky, who had the idea for radio astronomy. In an E-moment in December of 1932, Jansky realised that radio noise had a sidereal component, which meant that it was coming from outer space. He wrote an excited letter home on December 21 about this. Both I- and E-moments are lost — a distressing combination.

Appendix D. Lost moments (no dates could be found)

1952, **Bubble Chamber**: Donald Glaser made his first bubble chamber out of ether sealed in small glass ampoules. Cooled suddenly, they could record the tracks of atomic disintegration. He is said to have reached the idea while watching bubbles in a glass of beer. Dr Glaser told the author that he could not find amidst his old notes a record of the date.

1990, **World-Wide Web**: At CERN laboratories in Geneva, Tim Berners-Lee invented the world-wide web 'by' Christmas 1990. It was a gradual, step-by-step business: 'Inventing the world-wide web involved my growing realisation that there was a power in arranging ideas in an unconstrained, web-like way.' People would ask him if there had been some memorable moment when it all came together, and, as he explained in his book, 'They are frustrated when I tell them there was no 'Eureka' moment ...' (Weaving the Web, Tim Berners-Lee, 1990) Thus, there is no anniversary date for this tremendous modern invention.

APPENDIX E

Marginal cases

COULD THE E-GROUPS HAVE BEEN otherwise? Not everyone has agreed upon our time for Edison (appendix A). Other alterations considered were:

- Anton Lavoisier studied combustion, and on November 1st, 1772, confided a sealed note into the hands of the secretary of the French Academy, which said:
  'About a week ago I discovered that sulphur, in burning, far from losing weight, on the contrary gains it; ... This discovery, which I have established by experiments which I regard as decisive, has led me to think that what is observed in the combustion of sulphur and phosphorus may well take place in the case of all substances that gain in weight by combustion and calcination ...'
  This suggests an E-moment, but a historian of chemistry commented: 'No-one has ever explained what the experiment was that Lavoisier performed.'[3]
  Lavoisier's Q+S score is (4,2).

- HANS CHRISTIAN ØRSTED. While preparing for an evening lecture on 21 April 1820, on electricity, Ørsted noticed a compass needle that was being deflected from magnetic north, every time that the electric current from the battery he was using was switched on and off. This deflection convinced him that magnetic fields radiate from all sides of a wire carrying an electric current, just as light and heat do. It confirmed for him a direct relationship between electricity and magnetism.[4] At the time of discovery, Ørsted did not suggest any satisfactory explanation of the phenomenon, and it was only several months later that he began more intensive investigations.

EUREKA

- SAMUEL MORSE, an art lecturer of New York, made the electric telegraph work. In 1832, he was on board a sailing ship bound from le Havre to New York, where he watched a demonstration by Charles Jackson the Boston physician on electro-magnetism. It may then have dawned upon Morse how electric telegraphy could work, as he watched the demonstration of how an electro-magnet could be switched on and off.[5] Morse's career as a leading American leading artist terminated, as he crossed over from art to science, and a few years later assembled his telegraph design. Morse's Q+S score is (0,5).

- There is no reliable birth time for Max Planck, but there could be an E-moment: 'Planck most probably discovered his law in the early evening of Sunday, October 7th'.[6] [7]

APPENDIX F

Aspect-power in the harmogram

THE 'HARMOGRAM' PROGRAM DEPICTS THE flow of celestial energy (so to speak) for a given harmonic over a chosen time-period. To construct it, one models the shape of a celestial aspect over time, as it comes and goes, by a mathematical curve. A 'normal' Gaussian curve may be optimal, while a cosine curve is simpler to construct, and has a definite orb. Whatever forms are used, it is convenient if they enclose the same area over the designated orb. This means that, as a Harmogram depicts the coming and going of many aspects of a given harmonic and sums them, the average value of its 'aspect power' is the same as that of the aspect-score, and proportional to the orb chosen.

A simpler version of the Harmogram program (composed by M.O.) is based on a triangle-model of orb, whereby the force of a celestial aspect peaks at exactitude and decreases linearly to zero at the maximum orb, on either side. The program sums the relevant aspects for each moment in proportion to their distance from exactitude, so that

$$\text{Aspect Power} = \sum 2(A_{max} - A)/A_{max}$$

where A is orb and A_{max} is the maximum allowed orb. Thus, the triangle model has 'aspect power' proportional to distance from exactitude.

We usually set the 'base-orb' to twelve degrees, to give an orb of $(12 \div n)°$ for the nth harmonic, in accord with Addey's suggestion, but clearly this can be larger. If the conjunction is included, then the mean score would be three for any harmonic, for a base-orb of twelve, and so excluding the conjunction (as has here been done) slightly reduces that mean value.

We have here used a Gaussian or 'normal' curve in place of a triangle, to give the strength of each celestial aspect, as makes for a smoother curve. There is a problem in that it has no natural orb, as this curve extends out

EUREKA

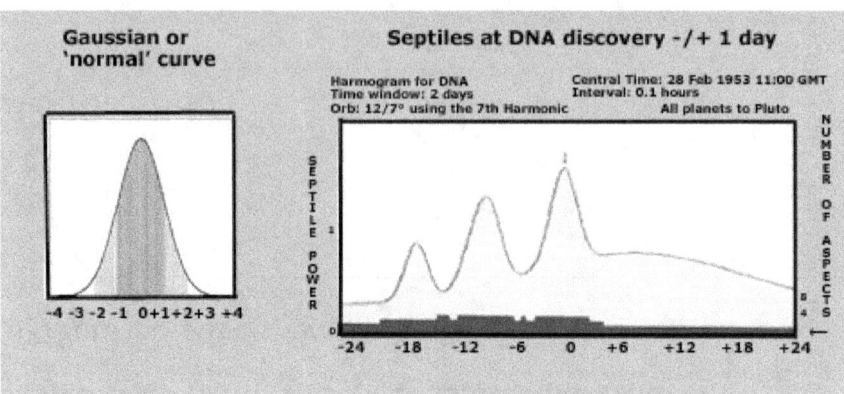

Figure F.1: Gaussian or 'normal' curve septiles at DNA discovery ±1 day. (Compare figure 12.12 for this DNA harmogram over an eight-day timespan.)

indefinitely, which M.O. solved by slicing a thin sliver off the base of the Gaussian curve, to give it a finite orb.

We conclude by showing the 7th-harmonic harmogram of the day of DNA's discovery by James Watson (figure F.1, certainly the finest dated E-moment of the twentieth century. One sees the smooth curves — peaking at the very instant of realisation — which result from using the Gaussian-curve to model the shape of a celestial aspect.

The Harmogram program is available from the web. Go to: `http://www.astrovdm.com/harmogrm.htm`, for instructions on how to install and use. It is a DOS program. To access DOS, go to `Start | Run ...` and type `CMD` to get a DOS command prompt. Type `EXIT` to return to Windows.

APPENDIX G

The moments of mystic illumination

Well-known eureka experiences, a part as we have seen of scientific folklore, would appear not to be continuing much beyond the discovery of the DNA helix structure by Watson and Crick in 1953; or at any rate, the present enquiry has not found them. Although, admittedly, a celebrated mathematical Eureka-cry did ring out when Fermat's Last Theorem was solved at Cambridge, in 1993. If more recent ones have occurred, then perhaps they have not been greatly noticed. Whereas, in contrast, a veritable cascade of reported and dateable moments of mystical enlightenment seems to be happening in recent times; with dateable mystic moments prior to, say, 1950 being relatively rare.

Figure G.1: The US sci-fi novelist Philip K. Dick, a photo and a portrait. He described quite a prolonged mystical experience, which arrived for him on 16th March 1974, in his opus 'Valis,' of 1991.

Books are written about these inward and transcendental moments, and we gather that the individual concerned has understood something important: but what has been comprehended is not readily explicable. Let's say it is *gnosis* rather than *scientia* that is found in such moments, and it is not — at least, not yet — testable by apparatus: rather it is something about the universe or our relation to it that seems to have been learnt. In the early mystic moments we find a contact with the Deity to be affirmed, whereas in more recent ones that is questionable. I noticed more or less the same excess of quintiles and septiles turning up in these moments, as for the 'Eureka' collection. If the life-blood of science is repeatability, then there would be a value in such a replication — the *fourth* here reported,

Table G.1: The Dated Mystic Moments (n=17)

Index	Date	Time (UT)	Zone	Name
1	1654 Nov 23	22:30hrs	+00:00hrs	Blaise Pascal
2	1739 Jul 23 (NS)	19:00hrs	+05:00hrs	David Brainerd
3	1744 Apr 7	22:00hrs	+00:00hrs	Emanuel Swedenborg
4	1895 May 11	05:00hrs	+05:00hrs	Paul Tyler
5	1921 Mar 22	20:00hrs	+00:00hrs	Martinus
6	1926 Nov 24	15:00hrs	−05:30hrs	Sri Aurobindo
7	1953 Mar 22	00:00hrs	−05:30hrs	Rajneesh
8	1964 Jan 15	16:00hrs	−02:00hrs	Muz Murray
9	1968 Dec 2	15:00hrs	−05:30hrs	Thomas Merton
10	1971 Feb 8	19:00hrs	+00:00hrs	Edgar Mitchell
11	1972 Aug 15	12:00hrs	+04:00hrs	Victor Mansfield
12	1974 Mar 16	12:00hrs	+08:00hrs	Philip K. Dick
13	1998 Apr 8	11:00hrs	−01:00hrs	Hazel Courtney
14	1998 Oct 10	12:00hrs	+00:00hrs	Matthew de Looze
15	2000 Jun 19	05:00hrs	−01:00hrs	Dave Oshana
16	2003 Sep 30	14:30hrs	−01:00hrs	Manjir Samantha-Laughton
17	2007 Jul 8	14:30hrs	−01:00hrs	Yvonne Aburrow

that discerns a significant excess for septile aspects. Using the same theory of harmonic aspects from John Addey as per the Eureka study, this group showed a deficit in what we may call 'stable-structure aspects' namely the 4th, 6th and 8th harmonics.

While this is work in progress, here are 17 mystic moments so far located, mostly having some time-of-day information. Here are three descriptions of them:

- The French mathematician, Blaise Pascal, experienced a mystic illumination, on Nov 23rd 1654, 10.30 p.m. to midnight. He kept an amulet with the memory of this inscribed upon it, and that was discovered after his death as tucked into the fabric of his coat. On it was written:

    FEU
    Dieu d'Abraham. Dieu d'Isaac Dieu de Jacob
    Non des philosophes et des savans
    Certitude joye certitude, sentiment, veue joye paix
    Dieu de Jesus Christ

    Other mystical comments of a joyful nature were inscribed, e.g., 'My God do not leave me. This is eternal life.'

- While orbiting the Earth in 1971, astronaut Edgar Mitchell watched how 'the earth, moon, the sun, and the planets all went through the window every two minutes', which he found to be a 'powerful, powerful experience:'

> What I experienced was a grand epiphany accompanied by an exhilaration. From that moment on, my life would take a radically different course... I actually felt what has been described as an ecstasy of unity ... I perceived the universe as in some way conscious ... the restraints and boundaries of flesh and bone fell away.

This led him to found the Institute of Noetic Sciences (10). This happened two days after leaving the Moon and a day before landing back on Earth so we may place it at 8 February, 1971 (they left the Moon 6 Feb 19hrs UT, and landed in the Pacific Ocean 9 Feb 21hrs UT).

- The British-Asian Manjir Samantha-Laughton was sitting on an oak-tree branch, taking her dog for a walk, when suddenly

  > It was if I had been thrown into the very fabric of the universe and caught a glimpse of its secrets. Everything seemed to be made up of the most exquisite geometry and mathematics, which was at the same time both infinitely complex and ever so simple. And I suddenly understood how black holes work and it wasn't at all how we had previously theorised.

She is Indian, and explained to me how her experience was 'a blend of mystical and science — in me the two are combined without conflict,' adding: 'The exact date was 30th September 2003, I do believe ...It was around 2–3 pm in the afternoon.' Her two books *Punk Science* and *The Genius Groove* both describe this moment. She lectures at prestigious scientific events, so whatever it was she realised, about how black holes, etc., work, seems to be taken seriously.

The mystic experiences have something transcendental in common, in a manner that expands or develops the identity of the subject. Totting up the harmonics (using, as before, the 'Jigsaw' astrology-research program) and using the same procedure as for the Eureka project, gave:

	Quintiles	Septiles
Totals for the 17 charts	55.0	57.0
Average per chart expected:	02.2	02.4
Expected for 17 charts:	36.3	40.9
Percent Excess:	**51%**	**39%**

To remind you, the Eureka project found 23 E-moments, with a distinctive excess of fifth and seventh harmonics, 17% for quintiles and 51% for septiles. These very same aspects are turning up again in the mystic moment group, significant at around one in a thousand ($\chi^2 = 11$).

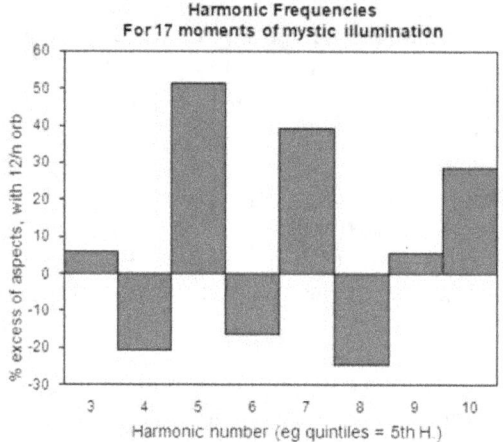

Figure G.2: Harmonic frequencies for 17 moments of mystic illumination.

The bar-chart shows a deficit of around 20% in the 4th, 6th and 8th harmonics. The 4th harmonic contains squares, the 6th contains sextiles and the 10th contains deciles. These are harmonics which we could describe as practical and functional in structural design, e.g., the one-sixth angle (6th harmonic) is the geometric basis for a beehive; while the number 8 is associated with the cube, and thus with 3-D space. These 'firm-structure' aspects have been somewhat avoided by the mystic moments: we may say that the heavens avoid such 'regular' celestial patterns at such times, and prefer the mould-breaking primes, 5 and 7. The bar-chart shows the large excess in 5th, 7th and 10th harmonics. The excess of 'tenth-harmonic' aspects appears as an 'overtone' of the 5th — somewhat as for the Invention-moments we saw how an excess of septiles 'echoed' in an excess of 14th-harmonic aspects.

Whatever it is that quintiles 'mean,' it is something absent from the public moments of technological achievement, of first-beginning, here referred to as 'Invention-moments.' In fact, they were substantially in deficit at such times, while being strongly present in both mystic and eureka moments. The present enquiry offers a further confirmation of the view that celestial aspects can be shown to have an effect — contrary to the view of the late Michel Gauquelin, who averred that this could not be done. Such effects are concordant with the harmonic theory of the late John Addey. Further discussion should perhaps wait until we see whether any further moments belonging to this group can be located.*

---

*'ISAR International Astrologer' Journal Summer 2011: N.K., 'Quintiles, Septiles and moments of Mystic Illumination'. See: http://www.astrozero.co.uk/articles/mystic_moments_kollerstrom.pdf.

References

1. For documents, see Marcello Pera, *The Ambiguous Frog*, Princeton UP 1992, trans. from Italian.

2. D. Guthrie, *Lord Lister*, 1947, p.59.

3. J .R.Partington, *History of Chemistry, III*, 1965 p.383.

4. *AOC*, p.230.

5. E. Larsen, *Ideas & Inventions*, 1960, p.249.

6. Abraham Pais, *Subtle is the Lord ... The Science and Life of Albert Einstein*, 1982, p.368.

7. See comments in Perkins, 'The Eureka Effect', p.171. There is no English-language biography of Planck.

Glossary

Addey orb: The orb of an aspect computed by $12° \div n$ where n is the harmonic number, giving e.g., for a trine $12° \div 3$ or $4°$.

Asimov scientist: Person cited in the *Biographical Encyclopaedia of Scientists and Inventors* by Isaac Asimov.

Aspect: A permitted angle measured in ecliptic longitude between two planets.

Applying aspect: An aspect before it reaches exactitude.

Aspect power: For a given harmonic at a moment in time, each aspect is scored by its distance from exactitude, larger as the orb is closer.

BDS: A reference work used to define the I-moment group: *Biographical Dictionary of Scientists: Engineers and Inventors*.

Biquintile: A $144°$ angle formed between two planets.

Biseptile: The angle $2 \times (360° \div 7) = 102°\,51'$.

Chi-square: A measure used to test significance in statistics, by comparing observed frequency against expected.

Cumulative frequency: Adds up the number of aspects in a group of charts as increasing with the orb used.

Deciles: Tenth-harmonic aspect: includes the $36°$ monodecile, $108°$ tridecile, etc.

Ecliptic: Plane of the Earth's orbit around the Sun, as a line through the heavens.

EUREKA

Eureka moment (E-moment): Time of achieving a notable insight.

Exactitude: The moment when an aspect has zero orb.

Expected frequency: The number of aspects for a given group computed on a chance basis.

Geocentric aspects: Aspects normally used by astrologers, measured from the Earth's centre.

Golden ratio: Value of ϕ where $1 + \phi : 1 = 1 : \phi$, approximately equal to 0.618.

Harmogram: Plot of aspect power against time.

Harmonic: A family of celestial aspects generated by an integer division of the circle.
For example, the fifth-harmonic includes quintiles and biquintiles, and the tenth-harmonic includes deciles and quintiles. The 'harmonic number' is that integer.

Histogram: A bar-chart frequency diagram.

Invention moment (I-moment): Time when a new invention first works.

Latitude: Celestial latitude is angular distance measured perpendicular to the ecliptic.

Longitude: Celestial longitude is angular distance along the ecliptic, from zero Aries, i.e. the Vernal Point.

Monoseptile: An aspect given by $360° \div 7 = 51°\,26'$.

Monte Carlo method: An empirical means of estimating the expected aspect frequency within a group.

Natal chart: Horoscope cast for a person's birth time.

Nodes (lunar): Points on the ecliptic where the plane of the lunar orbit cuts it.

Opposition: 180° angle between planets.

Orb: Angular distance of an aspect from its exactitude.

Planet: Includes Sun and Moon plus genuine planets Mercury to Pluto.

Ptolemaic aspects: The five celestial aspects recognised by antiquity: conjunction, opposition, trine, square and sextile.

Quintile: The fifth-harmonic aspects monoquintile (72°) and biquintile (144°).

Q+S: Quintiles and septiles.

Separating aspect: An aspect after its exactitude, with planets moving apart.

Sextile: 60° angle between planets.

Septile: The seventh-harmonic aspects: monoseptile (51°26′), biseptile (102°51′) and triseptile (154°17′).

Square: 90° angle between planets.

Trine: 120° angle between planets.

Triseptile: The angle of $3 \times (360° \div 7) = 154°17′$.

Table G.2: Astrological symbols.

	SIGNS		PLANETS		ASPECTS
♈	Aries	☉	Sun	☌	conjunction
♉	Taurus	☽	Moon	☍	opposition
♊	Gemini	☿	Mercury	△	trine
♋	Cancer	♀	Venus	□	square
♌	Leo	♂	Mars	✶	sextile
♍	Virgo	♃	Jupiter		
♎	Libra	♄	Saturn		
♏	Scorpio	♅	Uranus		
♐	Sagittarius	♆	Neptune		
♑	Capricorn	♇	Pluto		
♒	Aquarius	☊	North Node		
♓	Pisces	☋	South Node		

Select Bibliography

Books

Addey, J.M., *Harmonics in Astrology*, 1976.
Asimov, I. *Asimov's Biographical Encyclopaedia of Science and Technology*, 1966; 2nd Edition, 1975.
Aubrey's *Brief Lives*, Penguin, 1987.
A Biographical Dictionary of Scientists, Ed T. Williams, 1982.
Beveridge, W., *Seeds of Discovery*, 1965, 1980.
Brannigan, A. *The social basis of scientific discoveries* 1981.
Chandrasekhar, S. *Truth and Beauty*, 1986.
Cochrane, D., *Astrology for the 21st Century* 2002 Florida.
Culver, R. B. and Ianna, P. A., *The Gemini Syndrome, a scientific evaluation of astrology*, New York 1979, 1984.
Curry, P., *Prophecy and Power, Astrology in Early Modern England*, 1989.
Eysenck, H. J. and D. K. Nias, *Astrology, Science or Superstition?* 1982.
Gauquelin, M. & F., *The Gauquelin Book of Charts, Vol. 2 Men of Science* 1971; *The Gauquelin Book of American Charts*, ACS, CA 1982.
Gyhka, M. *The Geometry of Art and Life*, 1977.
Hadamard, J., *The Psychology of Invention in the Mathematical Field*, Princeton New Jersey 1945.
Hughes, T., *American Genesis: A Century of Invention and Technological Enthusiasm*, 1870–1970, Viking 1989.
Inglis, B. *The Unknown Guest, The Mystery of Intuition* 1987.
Kepler, J., *Kepler's Astrology Excerpts*, selected and translated by Ken Negus, Princeton NJ 1987.
— *De Fundamentis Astrologiae Certioribus*, Prague 1601, Trans. Bruce
— Brackenridge and Mary Ann Rossi as 'On the More Certain Fundamen-

tals of Astrology', in *Proc. Amer. Phil. Soc.* 1979,123.
— The Harmony of the World, (*Harmonices Mundi* 1618), 1997, Philadelphia, trans. E. Aiton, A .Duncan & J. Field.
Koestler, A., *The Act of Creation*, 1964.
Kollerstrom and O'Neill, *The Eureka Effect*, 1996.
Kuhn,T., *The Structure of Scientific Revolutions*, 1962.
Perkins, D., *The Eureka Effect*, 2001. Pottenger, Ed., *Astrological Research Methods, an ISAR Anthology*, (CA) 1995.
Rodden, L., *Astro-Data, Vols. I–V*, 1990 US.
Roberts, R., *Serendipity, Accidental Discoveries in Science*, New York 1989.
Schwartz, J., *The Creative Moment, How science made itself alien to modern Culture*, 1992.
Selye, H., *From Dream to Discovery, On Being a Scientist*, 1964.
Simonton, D. K., *Scientific Genius, A Psychology of Science*, CUP 1988.
Tarnas, R. *Prometheus, the Awakener* Woodstock, Spring Pubs., 1995.
Troinsky, E.H., *Das Horoskopos des Atom-Zeitalters*, Berlin, 1956.
Vickers, B. Ed, *Occult and Scientific Mentalities in the Renaissance*, 1984.
Weeks, M., *Discovery of the Elements, 4th Edition* 1939.
Young, A., *The Geometry of Meaning*, Robert Briggs Associates, CA, 1976.
Zukav, G., *The Dancing Wu Li Masters, an Overview of the New Physics*, 1979.

Articles

Brackenridge, Bruce, *Kepler's Fundamentals, The Scientific Basis of Astrology*, foreword to a new translation of Kepler's 'On the More Certain Fundamentals of Astrology' A. J., Winter 1980/1, pp.13–18.
Christianson, G. *The Night the Universe Changed Forever*, Griffith Observer, June 1997 4–10.
Donahue, W. 'Kepler's first thoughts on Planetary Motion' *Journal for the History of Astronomy*, 1993 xxiv.
Emsley, J., 'Mendeleev's Dream Table' *New Scientist*, March 7, 1985.
Field, J., 'A Lutheran Astrologer: Johannes Kepler', *Archive for History of Exact Sciences*, 31, 1984, pp.190–268.
Frisch, O., 'The discovery of Fission', *Physics Today*, Nov 1967, p.43.
Geison, G. and Secord, J, 'The Case of Optical Isomerism (on Pasteur)', *Isis*, 1988, 79, 6–36.
Graham, C. 'The Seventh Harmonic and Creative Artists', *Astrology Now*, June 1976.
Gruber, H. 'Aha Experiences', *History of Science*, 1981, xix, pp.41–59.
Kollerstrom, N., 'Pluto and Plutonium', *Astrological Journal*, Winter 1984, IV.

— & O'Neill, M., 'The Eureka Effect', *Astrological Journal*, 1988, 2, p.90–7;3, p.136.
— 'Kepler's Belief in Astrology' in *History and Astrology, Clio and Urania Confer*, Ed. Kitson, Annabella, 1989.
— 'Some Metallic Moments', *Astrology Quarterly*, Spring 1991, pp.33–43.
— & O'Neill, M., 'Invention-Moments and Aspects to Uranus', *Correlation*, Dec. 1992, pp.11–23.
— 'Investigating Aspects', Pottenger, 1995, pp.287–302.
— 'Kepler's Chart', *Astrological Journal*, Nov 1996, 371–7.
— 'The Moment of Invention — some anniversary dates', *Inventor's World*, Spring 1996 pp.14–15e
— 'Eureka Moments - anniversaries that shaped history', *Ibid*, Autumn 1998.
Maiman T. 'From Laser one to M1' by Hughes electro-optics, Malibu; also interview with Maiman in 'Lasers and Applications', pp.85–90, CA, May 1985.
Muller, Prof. A., 'Comments on Astronomical and Statistical Problems with Astrological Aspects', *Astro-Psychological Problems*, May 1986, pp.14–16.
Platt, and Baker, 'The Relation of Scientific hunch to Research', *Journal Chemical Education*, 1931, Vol.8.
Shanks, T. 'Astro-Psychological Problems', 1985, Ed. F. Gauquelin.
Shockley, W., 'The Path to the Conception of the Junction Transistor', IEEE Transactions on Electron Devices, 1984, 11.
Tarnas, R., 'Uranus and Prometheus', *Astrological Journal*, Sept 1989, pp.243–250 and June 1990 pp.150–156.
— 'The Western Mind at the Threshold', *Astrological Journal*, July 1991 pp.226–232.
Taylor, B., 'The Discovery of Pluto' in *Orpheus* Ed. Suzi Harvey, 2000.
Wotiz, J. and Rudofsky, S. 'Kekulé's Dreams: fact or fiction?', *Chemistry in Britain*, Aug 1984, 720–3.
Wright, P. 'Technology and the Uranus Cycle' *Astrological Journal, Winter 1983*.

On Inventions

The Biographical Dictionary of Scientists: Engineers and Inventors, *Ed. Abbott, 1985*.
Brown, G., The Guinness History of Inventions, *1996*.
Bruce, R., 'Bell', Alexander Graham Bell and the Conquest of Solitude *1973*. Burke, J., *Connections 1978*.
— The Day the Universe Changed, *1985*.

Clarke, R., The Scientific Breakthrough-the impact of modern invention, *1974*.

Flatow, I. They all Laughed ... From light bulb to Laser, the Fascinating Stories behind the Great Inventions that have Changed our Lives, *1992*.

Friedel, R. and P. Israel, Edison's Electric Light *N.J.*

Hatfield, H. S., The Inventor and His World.

Hazen, R. M., Superconductors: The Breakthrough, *1988*.

Heyn, E. V., The Fire of Genius, *New York 1976*.

Hodges, A., The Enigma of Intelligence *1987*.

Jewkes, Sawers and Stillermen, The Sources of Invention, *1958, 1969*.

Koch, W. E., The Creative Engineer — the Art of Inventing *1978*.

Larsen, E., Ideas and Inventions, *1960*.

Reader's Digest Association, The Inventions that Changed the World, *1982*.

Robertson, P., The Shell Book of Firsts *1974*.

Index

Ørsted, Hans, 41

Adams, John Couch, 287–290, *see* Neptune
Addey, John, 22–24, 49, 59, 77, 118
    orbs, 24, 50, 117, 302
    problem with orbs, 25
Airy, George
    Neptune, 90
Aldebaran-Antares axis, 59, 199, 207, 292–295
American lead in I-moments, 135
Anaesthetic, 163–167
Andromeda Galaxy, *see* Hubble
Applying vs separating aspects, 122, 224
Archimedes' E-moment, 1, 35
Asimov, Isaac, 33, 66, 68–71, 131–132, 315–319
Aspects
    effective, 24
    expected frequency, 72, 322
    Ptolemaic, *see* Ptolemaic aspects
Atom Bomb, 200

Atom split, 135
Atomic Clock, 205
Aubrey, John, 26, 286

Bacon, Francis, 13
Baconian method, 13–14
Baird, John Logie, 7, 135, 183
Balloon, 149–157
Banting, Frederick, 182
Barometer, 145
Bell, Alexander, 71, 169–170, 202
Besso, 3
Bethe, Hans, 75, 206
Birth data, 25–27, 315
Bohr, Neils, 107
Brahe, Tycho, 65, 262–265, 284
Brengger, Georg, 265

Chain, Ernst, *see* Penicillin
Chi-square significance test, 73, 321–324
Chu, Ching-Wu, *see* Superconductor
Churchill, Winston, 59
Cockcroft, John, *see* Particle Accelerator

353

Cockerell, Christopher,
  see Hovercraft
Computer, 202–205
  Algorithm, see Turing
  stored program,
    see Williams
  World's first, see Flowers
Control group, 321–322
Copernicus, Nicolaus, 74
Crick, Francis & DNA, 276–278
Crookes, William, 94
Curie, Madame as non-Eureka, 134

D'Alibard, Thomas, see Lightning conductor
Darwin, Charles, 86–89, 254, 290–293
Data collecting, 23–27
Data collections, 63, 68–71, 77
Davy, Humphrey, 66
  potassium, 84
de Fermat, Pierre, see Fermat's Last Theorem
de Medici, Cosimo II, 234
Decile aspects, at DNA discovery, 279–280
Descartes, 299–305
Diabetes, see Banting
Dictionary of Scientific Biography, 39
DNA E-moment, 68, 279, 300
  Crick, see Crick, Francis & DNA
  Watson, see Watson, James & DNA
Dynamo, 162, see Faraday

E-moments
  original list, 327
  undated, 112, 329
Edison
  birthdata, 315
  electric light, 123, 132–134, 172–175
  inventor, 217
  phonograph, 7, 66, 170–172
Einstein, 3, 66, 243–250, 252–255
Electric Light, 172, see Edison
Electric Motor, 160, see Faraday
Eureka Effect, 27, 69
Eureka publication by N.K. and Mike O'Neill, 69
Eureka publication by N.K. and Mike O'Neill, 27, 141
Eureka scientists, 65, 315
Eureka vs Invention moments, 7, 130
Expected frequencies, 72, 118, 224, 321–325

Faraday, Michael
  electric current induction, 66
Fermat's Last Theorem, 110–112
Fermi, Enrico, 67, 106
Fission, nuclear, 106
Flamsteed, John, 240
Fleming, Alexander
  E-moment, see Penicillin
Flight
  jet plane, see Jet Plane
  powered, see Powered Flight
Florey, Howard, see Penicillin
Flowers, Tommy, 203
Foucault, Jean Bernard, 167

Franklin, Benjamin, 6, *see* Lightning conductor
   personality, 241
Frisch, Otto, 106

Galileo
   his beliefs, 10
   his birthdata, 315
   moons of Jupiter, 64, 65, 233–234
Galle, Johann
   Neptune discovery, 66
Gene Therapy, 215
Giza Pyramids, 56
Goddard, Robert, *see* Rocket
Golden ratio, *see* Quintiles
Gruber, Howard, 270
Guglielmo Marconi, *see* Radio Station

H-bomb, *see* Thermonuclear Device
Hahn, Otto, 67, 106–107, 273
Halley, Edmund, 74
Hamilton, William
   and quaternions, 266
   and Wordsworth, 310
   E-moment, 77
Harmogram concept, 261, 337–338
   10th-H (DNA), 278
   2nd-H (computer), 205
   3rd-H (Kepler), 84
   3rd-H (jet plane), 196
   5th-H (Kepler), 264
   7th-H, 96–98, 263, 278, 296
   design of, 337
   using two harmonics, 78, 80

Harmogram concept 14th-H (quaternions), 269
Harmonic analyses, 49
   of E-group, 118
   of I-group, 223
   of mystic moments, 339
Harmonice Mundi (Kepler), 12, 252–256, 305
Heisenberg, Werner, 4
   matrix mechanics, 44, 45, 67
Helicopter, 192
Helium, 93–95, *see* Ramsay
Heptagon, 58
Herschel, William
   and Uranus cycle, 252
   music & philosophy, 231
   Uranus discovery, 75, 131
Heureka, 1
Hiroshima, 199, 272
Hologram, 211–212
Hooke, Robert, 216
Hovercraft, 208
Hubble, Edwin
   E-moment, 8, 42–43, 63, 67
Huxley, Thomas Henry, 293–295

I-moments list, 135
Illusory versus real E-moments, 5–6, 134
Insulin, 182, *see* Banting
Invention moments, 129–139, 143
Inventors birth data, 319

Jenner, Edward, *see* Vaccination
Jet Plane, 194–195
Jigsaw astro-research program, 222

355

Joliot-Curie, Frederic, 75
Jungk, Robert,
    see Thermonuclear Device

Kekulé myth, 5–6
Kepler
    and Brahe, 262–265, 284–285
    and Newton, 285
    E-moments, 19–20, 82–84
    ellipse discovery, 253
    his astrology, 11–12, 49–52, 255–257, 263, 265
    his chart, 12, 264
    Mysterium Cosmographicum, 20
    weather, see Brengger
Koestler, Arthur, 9, 33–43
Kosmos, 309
Kuhn, Thomas
    paradigms, 43
    X-rays, 96
    youth & age, 45

Laser
    construction by T. Maiman, 209–211
    idea by C. Townes, 7, 38, 107–109
Leibniz, Baron Gottfried, 253, 283
Leith, Dr Emmet, see Hologram
Leverrier, Urban
    Neptune, 90
Lightning conductor, 146–149
Loewi, Otto, 67
Lost E-moments, 329–332
Lowell, Percival, 102–106

Lunar Landing Module, 214
Lyons, Dr Harold, see Atomic Clock

Maiman, Theodore, 8, 209, 301
Mandelbrot Set, 76
Mandelbrot, Benoit, see Mandelbrot Set
Manhattan Project, see Thermonuclear Device
Mathematical E-moments, 6, 71, 76–80, 110–112
Meitner, Lise, 67, 106
Mendeleef, Dmitri, 3, 66, 92, see Periodic Table
Meyer, Lother, see Mendeleef
Microscope, see Hooke
Mitchell, Edgar, 340
Monk of Lisbon, 156
Morton, William,
    see Anaesthetic
Mystic moments, 339–342

Neptune
    and anaesthesia, 164–167
    discovery, 66, 90–92, 287–290
    orbit-period, 251–252
    seen by Galileo, 92
Nerve transmission, see Loewi
Newton & Kepler, see Kepler
Newton, Isaac, 235–240
    and apple, 5, 46, 235–240
    Optics, 57
Nicolle, Charles, 47
Noetic Science, see Mitchell
Non-Eureka group, 74–76, 317
Nuclear power
    and Szilard, 81–82

chain reaction, *see* Szilard,
  see also Fermi
  chart, 198–199
  fission, *see* Meitner, 199–353
  Hiroshima, 200, 207
Nuclear Reactor, 197

O'Neill, Mike, 22–24, 72
  graphs, 121, 224
  Harmogram design, 337–338
Oersted, Hans, 41, *see* Ørsted,
  Hans
Orbs, Addey's
  graphs, 121
  of higher harmonics, 118
  of Uranus-aspects, 222
  suggestion, 22–27
Oxygen, *see* Priestley

Périer, Florin, *see* Barometer
Particle Accelerator, 187
Pascal, Blaise, 340
Pasteur, Louis, 39–40, 112
Pauli, Wolfgang, 44
Pendulum, 167
Penicillin
  E-moment by Fleming, 67,
    98–102
  production by Florey
    & Chain, 192–193
Pentagrams & pentagons,
  see Quintiles
Periodic Table, *see* Mendeleef
Phonograph, 170, *see* Edison
Pluto, 102–106
  and fission, 198
  and plutonium, 271–274
  planet's discovery,
    see Tombaugh

Plutonium, 28–30, 271–276
Pocket watch, *see* Hooke
Poincaré E-moment, 6
Potassium, *see* Davy
Powered Flight, 177–181
Priestley, Joseph, 74
Ptolemaic aspects, 118, 137

Quantum mechanics, *see* Heisenberg
Quaternions, 266–270, *see* Hamilton
Quintiles, 21, 22, 28–30, 49–50
  and Einstein, 245–250
  and Golden ratio, 54, 279–280
  in Kepler's chart, 264
  of plutonium, 271

Röntgen, Wilhelm, *see* X-rays
Radar, 190–191, *see* Watson-Watt
Radio Station, 176
Ramsay, William, 66
Relativity Theory, *see* Einstein
Rocket, 186
Roentgen, Wilhelm, *see* X-rays

Sagan, Carl, 19, 75
Samantha-Laughton, Manjir, 341
Seaborg, Glenn, 28, 271–275
Septiles, 21, 56–59, 123–125
  in I-moments, 225
  in mystic moments, 342
Seventh-harmonic overtones, 118–119, 300
Shockley, William, *see* Transistor
Sikorsky, Igor, *see* Helicopter
Smallpox, *see* Vaccination

Solenoid, 161, see Faraday
Sputnik, 208
Stove at Ulm, 302
Superconductor, 213
Surgery, 163
Synastry, 283
Szilard, Leo, 67, 81–82, 197–199

Tarnas, Richard, 232–234
   on Freud, 253
Telephone, 168–169, see Bell
Television, 183–185, see Baird
Tesla, Nicola, 112–113, 133–135, 216–217
   birth data, 316, 317
The Lunar Nodes, 248
Thermonuclear Device, 206
Tombaugh, Clyde
   E-moment (Pluto), 67, 103
Torricelli, Evangelista, see Barometer
Townes, Charles, 7–8, 67, 107–109
Transformer, 161, see Faraday
Transistor, 201
Transits
   concept, 22
   of Darwin, 254
   of Einstein, 244–249
   of Montgolfier, 152–156
   of Newton, 251, 254–255
Trines
   and gene therapy, 216
   and jet plane, 195–197
   excess in I-moments, 223–224
Turing, Alan, 203–204
Twelve, number, 306–307

Upatneiks, Juris, see Hologram
Uranus
   cycle, 231–233, 252
   discovery of, 75, 131
   major aspects, 137–139, 222
   transits, 231–232

Vaccination, 158
Vitruvius, 1

Wallace, Alfred, 86–89, 290–292
Wallace, William, 124
Walton, Ernest, see Particle Accelerator
Watson, James & DNA, 3, 45, 276–278, 296
Watson-Watt, Robert, 132, 190–191
Whittle, Frank, see Jet Plane
Wiles, Andrew, see Fermat's Last Theorem
Williams, Frederick, 203
Wright Brothers, 178, see Powered Flight

X-rays, 66, 95–98, 175

www.ingramcontent.com/pod-product-compliance
Lightning Source LLC
Chambersburg PA
CBHW060550230426
43670CB00011B/1768